PHYSICS OF,
AND SCIENCE WITH,
THE X-RAY
FREE-ELECTRON
LASER

Related Titles from AIP Conference Proceedings

468   Nonlinear and Collective Phenomena in Beam Physics — 1998 Workshop:
International Committee on Future Accelerators
Edited by Swapan Chattopadhyay, Max Cornacchia, and Claudio Pellegrini, April 1999,
1-56396-862-2

413   Towards X-Ray Free Electron Lasers: Workshop on Single Pass, High Gain FELs
Starting from Noise, Aiming at Coherent X-Rays
Edited by R. Bonifacio and William A. Barletta, December 1997, 1-56396-744-8

395   Nonlinear and Collective Phenomena in Beam Physics – ICFA
Edited by Swapan Chattopadhyay, Massimo Cornacchia, and Claudio Pellegrini, July 1997,
1-56396-668-9

344   Nonlinear Dynamics in Particle Acclerators: Theory and Experiments
Edited by Swapan Chattopadhyay, Max Cornacchia, and Claudio Pellegini, September 1995,
1-56396-446-5

To learn more about these titles, or the AIP Conference Proceedings Series, please visit the
webpage **http://www.aip.org/catalog/aboutconf.html**

PHYSICS OF, AND SCIENCE WITH, THE X-RAY FREE-ELECTRON LASER

19th Advanced ICFA Beam Dynamics Workshop

Arcidosso, Italy   10–15 September 2000

EDITORS
S. Chattopadhyay
Jefferson Lab, Newport News, Virginia
M. Cornacchia
I. Lindau
Stanford Linear Accelerator Center, Stanford, California
C. Pellegrini
University of California, Los Angeles, California

Melville, New York, 2001
AIP CONFERENCE PROCEEDINGS ■ VOLUME 581

Editors:

Swapan Chattopadhyay
Jefferson Lab
Accelerator Division, MS 7A
12000 Jefferson Avenue
Newport News, VA 23606
USA

E-mail: cornacchia@ssrl.slac.stanford.edu

Max Cornacchia
Stanford Linear Accelerator Center
Stanford Synchrotron Radiation Lab
MS 69, P.O. Box 4349
Stanford, CA 94309-0210
USA

E-mail: cornacchia@ssrl.slac.stanford.edu

Ingolf Lindau
Stanford Linear Accelerator Center
Stanford Synchrotron Radiation Lab
MS 69, P.O. Box 4349
Stanford, CA 94309-0210
USA

E-mail: lindau@ssrl.slac.stanford.edu

Claudio Pellegrini
Department of Physics and Astronomy
University of California, Los Angeles
405 Hilgard Avenue, Box 951547
Los Angeles, CA 90095-1547
USA

E-mail: pellegrini@physics.ucla.edu

L.C. Catalog Card No. 2001092469
ISBN 0-7354-0022-9
ISSN 0094-243X
Printed in the United States of America

CONTENTS

Preface vii
Welcoming Address ix
    Mayor of Arcidosso

Summary of the Activity of the Group I: Physics and Technology of
the XFEL 1
    A. Renieri
Workshop Report: Science with the XFEL 6
    M. Sutton
Stability Evaluation of Femtosecond S-Band Linac with Photocathode
RF Gun 11
    M. Uesaka, T. Watanabe, T. Kobayashi, T. Ueda, K. Yoshii, K. Kinoshita,
    N. Hafz, H. Okuda, R. G. Hemker, and K. Nakajima
Beam Collimation for LCLS Beam Emittance and Charge Control 23
    C. B. Schroeder, H.-D. Nuhn, and C. Pellegrini
Wake Fields Effects due to Surface Roughness in a Circular Pipe 33
    M. Angelici, F. Frezza, A. Mostacci, and L. Palumbo
Plasma-Based Studies on 4th Generation Light Sources 45
    R. W. Lee, H. A. Baldis, R. C. Cauble, O. L. Landen, J. S. Wark, A. Ng,
    S. J. Rose, C. Lewis, D. Riley, J.-C. Gauthier, and P. Audebert
Effects of Channel Roughness on Beam Energy Spread 59
    A. V. Agafonov and A. N. Lebedev
Field Characteristics of Novel Hybrid Undulators Adequate to LCLS
Requirements: Simulation Results 73
    A. A. Varfolomeev and S. V. Tolmachev
Numerical Study of SASE Mode FEL Evolution Starting from Noise
with Subsequently Induced Slowing Down of Optical Pulse 78
    A. A. Varfolomeev and T. V. Yarovoi
Velocity Bunching in Photo-Injectors 87
    L. Serafini and M. Ferrario
Effects of Statistical Roughness on the Propagation of
Electromagnetic Fields in a Circular Waveguide 107
    E. Di Liberto, F. Frezza, and L. Palumbo
A Possible Experiment at LEUTL to Characterize Surface Roughness
Wakefield Effects 118
    S. G. Biedron, G. Dattoli, W. M. Fawley, H. P. Freund, Z. Huang,
    J. W. Lewellen, S. V. Milton, and H.-D. Nuhn
A Method for Increased Resolution of X-ray Images 124
    S. Monteiro
Predicted Performance of the LCLS X-ray Diagnostics 131
    E. Gluskin, P. Ilinski, and N. Vinokurov
Surface Roughness Impedance 141
    G. V. Stupakov
X-ray FEL with a meV Bandwidth 153
    E. L. Saldin, E. A. Schneidmiller, Y. V. Shvyd'ko, and M. V. Yurkov

Development of a Facility for Probing the Structural Dynamics of Materials with Femtosecond X-ray Pulses 162

B. Faatz, A. A. Fateev, J. Feldhaus, K. Floettmann, T. Tschentscher, I. Krzywinski, J. Pflueger, J. Rossbach, E. L. Saldin, E. A. Schneidmiller, and M. V. Yurkov

Diffraction Effects in the SASE FEL 169

E. L. Saldin, E. A. Schneidmiller, and M. V. Yurkov

Present Status of X-ray FELs 185

K.-J. Kim and Z. Huang

High-Gain Free-Electron Lasers and Harmonic Generation 203

S. G. Biedron, G. Dattoli, H. P. Freund, Z. Huang, S. V. Milton, H.-D. Nuhn, P. L. Ottaviani, and A. Renieri

An Optical Experiment towards the Analogic Simulation of the Betatronic Motion 211

A. Bazzani, P. Freguglia, L. Fronzoni, and G. Turchetti

Optimization of an X-ray SASE-FEL 221

C. Pellegrini, S. Reiche, J. Rosenzweig, C. Schroeder, A. Varfolomeev, S. Tolmachev, and H.-D. Nuhn

Participants List 229

Author Index 235

PREFACE

The 19th Advanced ICFA Beam Dynamics Workshop on "The Physics of, and Science with, X-ray Free-Electron Lasers" took place in Arcidosso, Italy, from the 10th to the 15th of September, 2000. The workshop was sponsored by the International Committee for future Accelerators, the U.S. Department of Energy, the University of California at Los Angeles, the Stanford Linear Accelerator Center, the Deutsches Electronen-Synchrotron, and the Lawrence Berkeley National Laboratory, together with local authorities of Tuscany, Grosseto, and Arcidosso. The Workshop was chaired by M. Cornacchia (SLAC), I. Lindau (SLAC/Lund University), and C. Pellegrini (UCLA). Seventy-five scientists attended the Workshop: 50 are involved in the physics and technology of accelerators, free-electron lasers and x-ray optics, and 25 pursue the scientific applications. There were plenary and parallel sessions and many lively discussions during and after the regular Workshop schedule.

Arcidosso is a medieval town in southern Tuscany, close to the city of Sienna. The meeting took place in the historically evocative setting of an 11th-century castle atop a hill dominating the nearby valley. The castle was restored in 1989, and preserves the atmosphere and ruggedness of medieval Europe.

There were two invited lectures on Monday, September 11, to open the subjects and two summary talks in the afternoon on Friday, September 15. All the other presentations were either informal or in the form of posters.

The group on "Physics and Technology of the XFEL," with introductory talks by Kwang-Je Kim (ANL) and Jamie Rosenzweig (UCLA), was coordinated by Alberto Renieri (ENEA-Frascati).

The group on "Science with the XFEL" was coordinated by Mark Sutton (McGill University) with introductory talks by Andreas Freund (ESRF) and Ingolf Lindau (SLAC/Lund University).

On behalf of the Organizing Committee, we would like to thank the sponsoring institutions for supporting the Workshop. We would also like to thank the Mayor, Ing. Attilio Marino, and the Administration of Arcidosso for hosting the meeting. We are grateful to the regional and provincial administrations, to the "Comunità Montana del Monte Amiata," to the representatives of local authorities and to the Cassa di Risparmio di Firenze for providing valuable support. Special thanks are due to the Workshop Coordinator, Ms. Melinda Laraneta, and the coordinator of the local organization, Sig. Gianfranco Nanni. The people of Arcidosso, as always, have contributed to creating a pleasant atmosphere with their welcome, warmth, and friendship. We thank the Program Committee, and the working group coordinators and speakers, who contributed to an exciting scientific program of presentations and discussions, and worked hard and with enthusiasm to implement it.

M. Cornacchia, I. Lindau, C. Pellegrini

Welcoming Address by the Mayor of Arcidosso

Ladies and gentlemen, good morning and welcome to all of you. first of all, as the Mayor of the city of Arcidosso and on behalf of the Municipality, I would like to greet the Authorities and to express my warmest thanks to all of the scholars and scientists who have gathered here in Arcidosso on the occasion of this congress devoted to the "Physics of, and Science with, the X-ray Free-Electron Laser".

A particular, respectful thank-you goes to the Italian Head of State, Carlo Azelio Ciampi, under whose auspices this important event takes place once again: the President of the Italian Republic wishes to highlight the scientific importance of the Conference and to recognize the value of the academic institutions represented here, together with the high intellectual profile of its participants.

I also wish to express my gratitude to the ICFA, which, by accepting to support this Conference, has contributed to increase the worldwide interest of the academic world on this important event.

I would like to start my brief speech by expressing my personal satisfaction in being here again, for the third time, to welcome all of you. Every time I have been granted the possibility to open such a Conference I feel proud to be the Mayor of this small town, and I am inclined to think that all this does not happen by chance.

In fact, a consolidated history of cultural traditions and of significant coincidences have contributed to bring to the national and international attention the town of Arcidosso, establishing its role as a place where high-profile scientific conferences take place.

However, and in so doing I hope not to offend anybody, I think that the most important of all these conferences, the most awaited one, remains this one organized by Profs. Cornacchia and Pellegrini. Prof. Cornacchia, a native of Arcidosso, has been able to combine the affection for his place of origin with a deep involvement in his field of studies, thus bringing to the attention of the world to Italy, and the town of Arcidosso, in particular.

To have all of you gathered here once again makes me extremely happy, and I want to share this happiness with all those who, in the last two years, have worked hard to organize and to prepare this conference.

Some of you have already been our guest in the past: thus, you have had the possibility of appreciating the simple though warm welcome that this small medieval Tuscan town is able to offer its visitors. Your participation to this conference contributes to put Arcidosso, though for a little while, at the center of the international attention.

To those of you who come to Arcidosso for the first time I want to address my warmest welcome, hoping you will appreciate the beauty and serenity of this part of the Amiata mountain, with its rich cultural heritage but, at the same time, ready to search for a new, modern identity. Back in the 8th century AD, Mount Amiata was dominated by the Benedictine monastery located in Abbadia San Salvatore, an important religious, social, economic and cultural center. Around the year 1000 the Aldobrandeschi family, of Longobard origins, played a prominent role in the area: the first historical documents mentioning the town derives from the Latin words "ARX" and "DOSSUM", meaning a fortified castle on the top of a rock.

This concerns the past. As for the present, our community, though aware of the cultural and spiritual values it has inherited from such a long historical tradition, is trying to find a prominent and important place for itself in the modern economic scenario, especially by promoting the tourist industry. We are in fact deeply convinced that we can offer many things to the tourists who can appreciate the natural beauties, the spirituality and the history of our territory.

In thanking again all the participants to this conference, I wish you a fruitful exchange on such important topics and I renew my thanks to all those who have participated to the organization of this conference. In particular, I wish to express my thanks to the Councilor for Cultural activities of the Municipality of Arcidosso, Mauro Nascari, and to Gianfranco Nanni, Aviano Bargagli, Melinda Laraneta (the workshop coordinator), and to all the sponsors who have helped make this possible: the Administration of the Tuscan Region, the Provincial Municipality of Grosseto, the Local Agency for Tourism, the Comunità Montana and the bank Cassa di Risparmio of Florence.

Again welcome to all of you and work well!

Attilio Marino
Mayor of Arcidosso

SUMMARY OF THE ACTIVITY OF THE GROUP I: PHYSICS AND TECHNOLOGY OF THE XFEL

Alberto Renieri* (Group Co-ordinator)

* ENEA, Divisione Fisica Applicata, Centro Ricerche Frascati, C.P. 65, 00044 Frascati, Rome Italy

INTRODUCTION

The main topics investigated in Group I (Physics and Technology of the XFEL) have been
A) Accelerator (photo-injector, beam instrumentation, etc.)
B) Collective effects and instabilities
C) SASE FEL process (physics, exotic schemes, etc.)
D) Undulator instrumentation.
The wide scientific and technological spectrum involved in suggested to articulate this Group in some more topical Sub-Groups. Namely it has been decided to divide the Group into 4 Sub-Groups following the topics reported above. The Co-ordinators of the 4 Sub-Groups have been the following,

A. T. Limberg (Accelerator)
B. L. Serafini (Collective effects and instabilities)
C. G. Dattoli (SASE FEL process)
D. J. Rosenzweig (Undulator instrumentation)

Large part of the work has been made with all the Group I members together, while some more specific analysis has been better and more deeply investigated in the specific Sub-Groups (namely Wednesday morning there were separate meetings).
Many contributions have been presented (31 oral contributions and 17 posters) and the discussions stimulated by the presentations have been effective and fruitful and their outputs provide an essential component of the Workshop.

CP581, *Physics of, and Science with, the X-Ray Free-Electron Laser,* edited by S. Chattopadhyay et al.
© 2001 American Institute of Physics 0-7354-0022-9/01/$18.00

The main issues considered have been,

- Injector (in particular emittance vs. charge)
- Pulse compression (from few hundreds fs to few tens fs)
- Wake fields (resistive wall, surface roughness, pipe discontinuities)
- System optimization (choice of the main parameters, general scaling laws for gain and saturation including wake fields effects)
- Harmonics and exotic schemes (shorter wavelengths, more coherence)
- X-rays and electron beam diagnostics (what it is really needed and what it is really possible to measure and with what precision)

In the following sections a short summary of the main conclusions derived during the workshop activity, and widely discussed in the plenary section held the last day of the workshop, is presented.

ACCELERATOR (SUB-GROUP A, CO-ORDINATED BY T. LIMBERG)

The general feeling about the main electron bunch parameters is that the "standard" request of

$$\text{charge} = 1 \text{ nC, normalised emittance} = 1 \text{ mm·mrad}$$

appears reasonable, namely this kind of performance has been already obtained (photo-cathode technology), but, up to now, this goal it is not yet routinely reproducible. This means that there is still a lot of work that need to be done. In particular it appears that the level of laser and cathode technology is quite adequate, while diagnostics systems need further important development. Furthermore it must be underlined that the emittance preservation is still an open question.

A deep discussion was focused on alternative parameter set options. Namely it could appear convenient to operate at lower charge (and with lower emittance to fit the FEL requests) in order to reduce, e.g., the wake fields problems. As a test case the option

$$\text{charge} = 0.2 \text{ nC, normalised emittance} = 0.6 \text{ mm·mrad}$$

has been considered. The feeling is that this configuration too is feasible as the 1 nC, 1 mm·mrad option, but no optimisation has been already done. This optimisation work has been indicated as a quite important activity to be done in the near future.

Perspective from two new kinds of guns have been discussed. Namely,

- Electrostatic + RF gun (Van der Wiel): it is based on present technology but it could be utilised for particular applications only (in particular the duty cycle is quite low)

- Plasma gun (J. Rosenzweig): this is a really new technology and its perspectives are quite appealing, but it is still at a very early stage of development

COLLECTIVE EFFECTS AND INSTABILITIES (SUB-GROUP B, CO-ORDINATED BY L. SERAFINI)

The more challenging issue appeared to be the enhancement of the electron bunch energy spread due to the wake fields generated by the beam pipe surface roughness. Four models have been considered ("Agafonov", "Palumbo-Di Liberto", "Stupakov", "Novok"). The outputs from these models do not agree, but it must be considered that they are based on different configuration assumptions and, moreover, they have different range of validity. This means that it was not so easy and straightforward a reliable comparison between them. The general feeling was that the effect can be dangerous and a reduction of the bunch charge (from 1 nC to 0.2 nC) could be quite useful. Strong recommendations for future work are listed below,
- The analytical models to predict realistic surface roughness conditions (in particular a realistic expect ratio) must be enhanced. In particular:
  - ➤ Agafanov model: it must be improved in order to be able to deal with long channels too,
  - ➤ Palumbo-Di Liberto model: it is quite important that the planned work program would be completed, i.e. longitudinal field components and wake potential derived from Green function model of random surface roughness.
- The capability of numerical simulation tools must be pushed to model a realistic randomly distributed surface roughness. In this way it would be possible to validate the analytical predictions.
- Experiments focused on the application of "controlled" surface roughness conditions to the beam pipe must be performed (the possibility of experiments in LEUTL an TTF has been discussed).

To conclude, it has been stressed that resistive wall and pipe discontinuities need further investigation (both simulations and experiments).

SASE FEL PROCESS (SUB-GROUP C, CO-ORDINATED BY DATTOLI)

One of the main problems appeared to be the enhancement of the energy spread due to the wake fields (see previous section). In particular the situation in higher harmonics operation must be carefully investigated. It has been stressed that experimental measurements will be available in short time, namely in one year in LEUTL and at the end of 2000 in TTF. In the mean time optimisation criteria in presence of collective effects and taking into account the scaling with the charge have to be developed. A particular recommendation come from LEUTL. Namely, as stressed before, LEUTL FEL will reach saturation soon and the colleagues in Argonne like to know from the community what it is really important to measure, in order to derive good scaling laws.

As to the HGHG (high gain harmonic generation) the status of art seems to be mature. In particular there is a good agreement between theory and experiments. This operation mode appears a good tool for:
- Smoothing X-ray beam
- Tunability down to 0.5 Å

In this field appears important to more deeply investigate,
- Self modulation
- Seeding
- Harmonic enhancement

Finally the two steps scheme for improving the X-ray beam quality (Rossbach) has been discussed and it was suggested to deeper investigate this topics.

UNDULATOR INSTRUMENTATION (SUB-GROUP D, CO-ORDINATED BY ROSENZWEIG)

A quite extensive analysis of the diagnostics needed for the correct handling of the electron beam and X-ray beam has been done. The main issues that have been single out are reported in the following two subsections.

The Electron Beam

In order to operate a SASE FEL correctly the electron beam must be matched to the focusing lattice and to the undulator axis and then an accurate diagnosis system must operate in a reliable way. The electron beam parameters to be monitored are the centroid, the transverse profile and the longitudinal characteristics. Namely,
a. For the centroid (μm resolution) the following devices can be utilised,
   - RF BPM (non destructive)
   - OTR (destructive)
   Two main questions arise,
   - Does OTR survive?
   - How does surface quality affect light?
   Both questions must be answered by experiments soon
b. For the transverse profile (β-matching),
   - Scintillators are unacceptable due to the saturation
   - OTR (they will survive at lower bright beam pre-tuning)
   The main question regards the feasibility of emittance measurement with the resolution of the order of 1 mm•mrad.
c. Longitudinal characteristics,
   - Pulse length must be pre-measured (with CTR (which is destructive), CSR and CDR)

- Time resolved slice measurements (emittance and energy spread) can be made with RF deflection "streak" that works better at high energy and low emittance.

The X-Ray Beam

The main X-ray beam parameters to be measured are the intensity (longitudinal profile), spectrum (as a function of λ and θ) and the temporal characteristics. The following questions arise, that need soon an answer
- Can we obtain information on spontaneous emission from coherent light signals?
- Can diagnostics devices survive at the X-rays and electron beam fluxes?

The proposed diagnostics rely on diamond, crystals, multilayers. X-ray flux is enormous, well above optical damage threshold. But damage by X-ray is quite different. The dominant mechanism for damage must be estimated by taking into account ionisation processes. Avalanche ionisation is unlikely for X-ray only, but materials directly impacted by electron beam may see quite large space charge fields and avalanche ionisation is very likely. One solution could be to steer the electron beam away. Experiments are needed.

CONCLUSIONS

As a final remark it is possible to say that, although the status of the art of X-ray SASE FEL appears now in a quite advanced position respect to only few years ago, a lot of work remains to be done. My personal feeling is that one of the main results of this Workshop is that the real important issues have been clearly singled out and the direction of the experimental and theoretical investigation has been well defined. We can expect now that in the following year the activity stimulated by this Workshop will give the requested answers to the scientific community involved in this field.

Workshop report: Science with the XFEL

Mark Sutton

Centre for the Physics of Materials, McGill University, Montreal Quebec, Canada H3A 2T8 email

This report summarizes the discussions help by Group II of the ICFA Advanced Beam Dynamics Workshop called *Future Light Sources, Physics of, and Science with, the X-ray Free-Electron Laser*. It starts with a discussion of several experiments which will benefit from the high intensity and the short pulses of the proposed XFEL beams. The choice of these experiments reflects the interests of the participants and the proposed sets of first experiments to be done with the XFEL. The second section discusses some of the instrumentation and optics issues that arise because of these experiments. The final section lists a set of topics that need active research and development in order to fully utilize these sources.

1. Experiments

The following experimental descriptions are purposefully brief and only intended to put the experimental requirements in perspective. More detailed descriptions can be found in the reports from SLAC (http://www-ssrl.slac.stanford.edu/lcls) and DESY (TESLA reports 1-3) (http://tesla.desy.de).

From the wide variety of experiments discussed it is clear that each experiment has different requirements. Meeting the strict requirements of all these experiments puts quite severe constraints on the XFEL. It is worth stressing, however, that given a set of specifications of a running XFEL many interesting experiments can still be designed and performed even if it doesn't meet all of these requirements. It is hoped that by listing the individual experiments below some idea of the requirements for each experiment can be assessed. At the moment, requirements are more qualitative then quantitative. One of the many things that still need to be done as the experiments become better defined is to properly quantify these requirements.

Atomic Physics: This title covers a variety of experiments on the basic interactions of short but intense x-ray pulses and atoms. Besides advancing our understanding of the interactions between EM waves and atoms, an understanding of a subset of this information is essential some of the other topics to be discussed as an understanding of the fundamental interactions will useful in the design of optical elements, sample geometries etc.

Proposed experiments include: forming multiply ionised core levels in atoms; studying

CP581, *Physics of, and Science with, the X-Ray Free-Electron Laser*, edited by S. Chattopadhyay et al.
© 2001 American Institute of Physics 0-7354-0022-9/01/$18.00

highly ionised atomic clusters; multiphoton ionisation and other non-linear effects. One exciting possibility is that these studies could lead to possible lasing of XFEL excited matter.

The first experiments to measure non-linear effects involve measuring various cross-sections as a function of energy density (intensity), an absorption cell will allow lower intensity levels to be measured and focusing will give access to higher energy densities. Other experiments would use wavelengths near various absorption edges to excite various decay modes and measure life times, again as a function of intensity. The properties of the XFEL pulse will be appropriate for many of these systems but many also have processes faster then 200 fs so they could use even shorter pulses. These experiments would use the full range of wavelengths 1.5 to 15 Angstroms (and probably more if given to them) in order to cover various absorption edges and processes. Also since the experiments are basically measuring cross-sections a measure or knowledge of the incident beam intensity is needed.

Plasmas and Warm Dense Matter: These experiments will use the high energy density of the beam to heat matter to extreme temperatures and pressures so that its properties can be measured (warm dense matter). With the energies deposited by the XFEL it is expected that states of matter can be created that span the range from conventional solid-state to plasma-like states. One way to study these states is by the resulting emission of lines as the material heats and cools. The experiments could take advantage of focusing techniques to reach higher energy densities. Another potential type of experiments is to split off a piece of the x-ray pulse, delay it, and then use it for Thompson scattering to study the structure and modes of the material as if evolves in time.

A related class of experiments, is to measure x-ray diffraction (Thompson scattering) off the short lived plasmas created by high intensity laser pulses synchronized to hit just before the XFEL x-ray pulse.

Femtosecond Chemistry: The underlying goal of chemistry is to understand the molecular dynamics of the atoms involved in a chemical reaction. The XFEL provides a way to follow the atomic positions during a reaction by using time resolved x-ray diffraction. The possibility also exists to study clusters and other nanostructures as they melt, freeze or vibrate. As well, other time resolved x-ray probes such as XANES or EXAFS would give interesting results.

A typical experiment consists of exciting a chemical reaction by a pump laser and measuring the x-ray diffraction pattern using a delayed x-ray pulse from the XFEL. Since these are primarily diffraction experiments the shorter XFEL wavelengths would tend to be used. To further increase count rates, the x-ray beams could be usefully focussed to smaller sizes.

A lower bound of about 8 fs (set by the time it takes H - OH to disassociate) exists on the time scales for chemical reaction and so although many useful measurements can be performed with 200 fs pulses, shorter pulses could advantageously be used. One route to these shorter pulses could be pulse slicing techniques.

Single Biological Molecules and Clusters: The ultimate goal of this class of experiments is to shine enough x-rays on a large biological molecule to measure its structure in a time less than the time it takes for its decomposition from the effects of the x-ray pulse. This would greatly increase the number of biological molecules whose structure could be studied by x-ray diffraction. In particular, there is a large number of important molecules which cannot currently be crystallized.

Current estimates give that with extreme focusing (100 nm spot) and short pulses (50 fs) such intensities could be reached with an XFEL. Even with intensity levels short of this level, important information could be obtained (even complete structures) from small clusters of molecules or by using information from repeated measurements.

This experiment would like state of the art focusing (or better) and since it is a diffraction experiment, would like to use the shorter wavelengths. Also, it would ideally like x-ray pulses closer to 1 fs and intensity levels up to 5×10^{13} photons per pulse, if ways could be found to push these XFEL characteristics.

Imaging/holography: Conventional imaging, holography and structural studies typically need information from many different orientations of the material under study. It is thus difficult to come up with schemes to take advantage of the short pulses of the XFEL. One such scheme is to perform holography by using the interference of the fluorescence signal emitted with an incident x-ray beam. By detecting the fluorescence signal in an angle resolved mode this it is possible to determine the local structure of the fluorescing species. Preliminary experiments have been down at existing synchrotrons and estimates show that there is enough signal to image small crystals in a single XFEL pulse. This opens the possibility of sub-picosecond time resolved measurements of local environments.

As proposed, this technique would be useful for K and L edges of materials in the many Kev energy ranges as these wave lengths are close to or slightly shorter than interatomics distances.

X-ray Intensity Fluctuation Spectroscopy (XIFS): A transversely coherent x-ray beam diffracted by a disordered medium produces a speckle pattern. XIFS is performed by measuring the time evolution of this speckle pattern and provides a direct measure of the time evolution of the system. It is one of the few techniques which can measure the fluctuations in an equilibrium system. By using a split pulse with varying time delays it should be possible to measure time constants in systems from picoseconds to many tens of nanoseconds or longer.

This technique is based on using the x-rays from an XFEL as a non-interacting probe of matter. Estimates using the ratio of scattering to absorption suggest that, with some attenuation, it will be possible to get enough scattering off an amorphous material made of low Z material to measure a speckle pattern in a single pulse. For more highly structured materials, like nanoclusters or disordered crystals, materials with higher Z are also accessible. The ratio of scattering to absorption is typically much improved for higher energies and the estimated intensities given for the third harmonic of XFEL would potentially also be beneficially used for this experiment. Furthermore, this is a high resolution diffraction measurement and issues of beam position could matter. Since it requires high angular resolution long sample to detector lengths would be convenient.

X-ray Free Electron Laser Physics: There are many interesting physics issues about the many processes involved in producing x-rays from an FEL. For instance: how to decrease pulse length, how to increase longitudinal coherence and/or intensity, how to chirp the beam and methods of harmonic generation to name a few. Instrumentation and time to study these issues needs to be allocated.

2. Instrumentation and Optics

The above experiments naturally lead to several required optical components and other techniques to handle the x-ray pulses.

Focusing and monochromators: Most (if not all) of the optical requirements for focusing and monochromators can be met by existing optics. However, the existing setups use materials that are not be compatible with the destructive power of XFEL beams (too high-Z). Thus work needs to done to make monochromators, mirrors, zone plates and diffraction gratings or multilayers out of low-Z materials which can withstand the intense XFEL beams. An possible alternative for optics, is to develop cheap optics that they can be moved or replaced for each XFEL pulse. Since the beam area is small only small displacements will be needed. The rapid movement of the optics could put severe constraints on the required motions.

Pulse techniques: A variety of techniques need to be developed for exploiting the pulsed nature of the XFEL. Methods of synchronizing the experiments to the x-ray pulse are needed. For example, some of the suggested experiments require synchronizing an optical laser relative to the x-ray pulse.

Techniques for pulse splitting and delay lines is another example needed for many of the experiments described above. Pulse stretching or shrinking techniques is possible if the x-ray pulse can be chirped. Another set of useful techniques would be pulse slicing techniques. These would produce shorter pulses and although they are at reduced intensities they would give higher time resolutions. Suggestions for many of these techniques have been proposed but work is needed to develop, test and demonstrate their feasibility.

Beam monitoring: Since many of the beam parameters, such as position and intensity fluctuate from pulse to pulse techniques to measure these need to be developed and tested. For some experiments post processing with this information would give the higher precision required.

Detectors: Since most photon counting detectors have dead times longer than the XFEL pulse lengths, they will not be useful. Techniques for energy resolving and spatial resolving detectors will be needed for many of the above experiments. Some mechanism for autocorrelating x-ray pulses or converting the time information into spatial distances are also required.

3. Research and development topics

In this section we list a set of research topics that will have an important impact on the usage of an XFEL. Research and development of many of the topics can and should begin now.

1. Better general understanding of how the XFEL radiation is different from conventional SR. For instance, the effects of FEL wake fields, near field effects at undulator output, how short pulses interact with optical elements (mono, crystal, gratings, focusing) and pulse to pulse intensity variations need to be understood. It is important to encourage existing, or soon to be operational FELs, to test these ideas as much as possible.
2. Development of FEL optical elements. Crystals, multilayers with new materials, zone-plates, refractive lens, plasma lens, liquid mirrors. For optics whose solution to the damage problem is repositioning after each shot, issues of precise positioning of elements need to be clarified.
3. Perform experiments on the effects of high power densities on materials. For instance, using existing high power laser facilities to experimentally simulate some aspects of such an environment.
4. Closely related is to work on the theory to help understand and predict these effects. One can develop, adapt and test basic theory and computer codes (which should be coupled to existing XFEL codes) with an emphasis on understanding properties and uses of proposed XFELs.
5. Work through in more details the layouts for specific experiments (optics, sample manipulation, detectors, data acquisition).
6. Begin studies of non-linear interactions using existing sources. Encourage theorists to work on these problems and make predictions.
7. Develop and test sub-picosecond time correlation and autocorrelation measurements. Work on synchronizing everything to appropriate fractions of the pulse length. Also work on pulse splitting techniques.
8. Develop detectors for a range of things, incident intensity and beam position monitors, large area detectors, high dynamic range detectors,
9. Continue to work on new ideas for possible experiments to extend the list above. Make general community aware of this new x-ray source, its properties and its potential.

Stability Evaluation of Femtosecond S-Band Linac with Photocathode RF Gun

Mitsuru Uesaka, Takahiro Watanabe, Tetsuya Kobayashi, Toru Ueda, Koji Yoshii, Kenichi Kinoshita, Nasr Hafz, Hiroyuki Okuda, Roy G. Hemker and Kazuhisa Nakajima

Nuclear Engineering Research Laboratory,
School of Engineering, University of Tokyo,
2-22 Shirakata-shirane, Tokai-Mura, Naka-Gun, Ibaraki 319-1188, Japan
E-mail : uesaka@tokai.t.u-tokyo.ac.jp

Abstract. Femtosecond S-band linac with the BNL/GUN IV of Nuclear Engineering Research Laboratory, University of Tokyo, has recorded 240fs(FWHM), 6 pai mm. mrad(rms), 350pC/bunch. Maximum charge per bunch at the exit of the injector and QE were 7nC and 1.5×10^{-4},respectively. After the cathode damage was found, re-measured emittance was 20 pai mm. mrad. Timing jitter between the femtosecond electron bunch and the drive laser for the injector was 330fs for the minutes and 1.9ps for hours. Detailed of the system are described, including the electron bunch diagnosis system. Further, the plasma gun work to produce 10fs, 20MeV electron single bunch without any timing jitter is also mentioned. Those date and information give important mile stones for development of future SASE-FEL including LCLS.

INTRODUCTION

Femtosecond Ultrafast Quantum Phenomena Research Facility was established at Nuclear Engineering Research Laboratory, University of Tokyo, in 1999. Research and application of femtosecond electron single bunches have been carried out at the S-band twin linacs, where the BNL/GUN IV photoinjector [1] has been installed in the newer 18MeV linac. Especially, the new 0.3TW100fs Ti:Sapphire laser is used to drive the photoinjector, to achive subpicosecond timing jitter ant to be applied for radiation chemistry work. As for electron bunch diagnosis, we use the femtosecond streak camera, coherent transition radiation(CTR) interferometer, far-infrared polychomator and fluctuation method. Furtermore, numerical analysis and experiment on the plasma gun to produce 10fs, 20MeV electron single bunch is under way using the 12TW50fs Ti:Sapphire laser. Updated results are described.

CP581, *Physics of, and Science with, the X-Ray Free-Electron Laser,* edited by S. Chattopadhyay et al.
© 2001 American Institute of Physics 0-7354-0022-9/01/$18.00

PERFORMANCE OF S-BAND RF PHOTOINJECTOR AND LINAC

The Femtosecond Ultrafast Quantum Phenomena Research Facility consists of the upgraded femtosecond S-band twin linacs, the 12TW50fs laser, the X-ray diffraction analysis devices, the X-ray CCD camera system, the X-ray photo-electron spectroscopy (XPS) device and the Fourier transform infra-red spectroscopy device (FTIR) as shown in Fig.1.

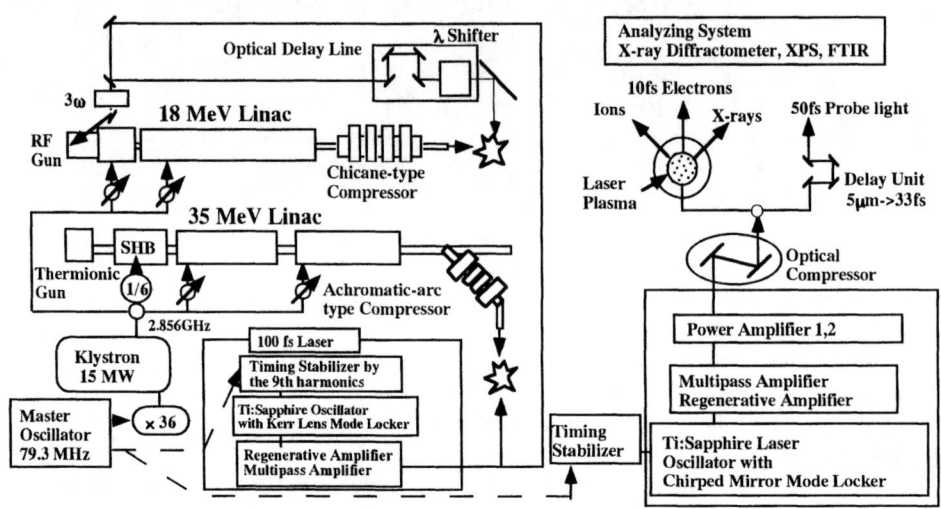

FIGURE1. Femtosecond Ultrafast Quantum Phenomena Research Facility.

Configuration of the S-band second linac with the laser photocathode RF gun, BNL/GUN-IV, is drawn in Fig.2. It consists of the gun, one 2m- long S-band accelerating tube, chicane-type magnetic bunch compressor and several diagnostic devices. The cathode material is copper.

Overall evaluation of the performance of the gun and linac was carried out by using the diagnostic devices such as the quad-scan, energy analyzer, Faraday cup and streak camera with 200fs resolution at FWHM (Hamamatsu FESCA-200). Measured beam energy, emittance, energy spread, pulse width and charge per bunch as a function the RF gun phase are described in details in ref. [2,3]. The emittance measurement was done after acceleration up to 16MeV and without using the chicane. The maximum energy of 3.33 MeV, lowest emittance of 6 pai mm.mrad, shortest pulse width of 10ps and maximum charge of 2nC were obtained at the different phases. Typical results of bunch shapes without and with the bunch compression are given in Fig.3. The 13ps bunch is compressed down to 440fs at

FWHM. After the calibration of the streak camera using the 86fs (FWHM) Ti:Sapphire laser (Spectra Physics, Tsunami), the error reduction gives the bunch length of 240fs(FWHM).

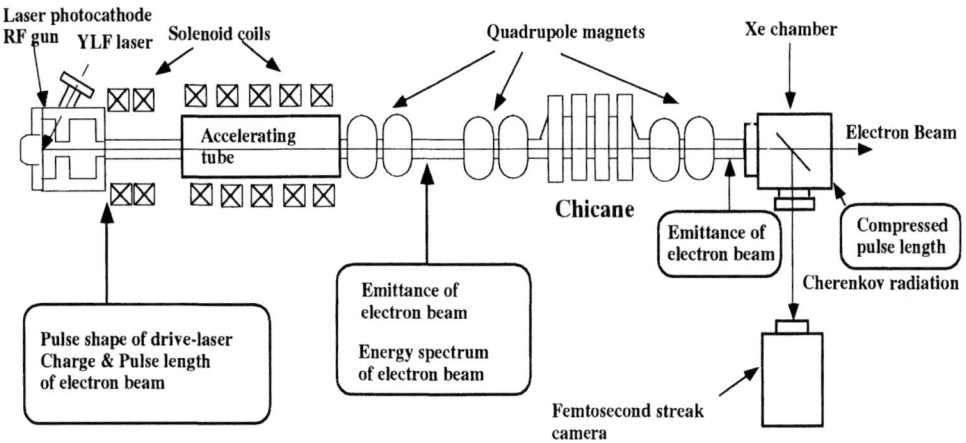

FIGURE2. S-band RF photoinjector and 18L linac.

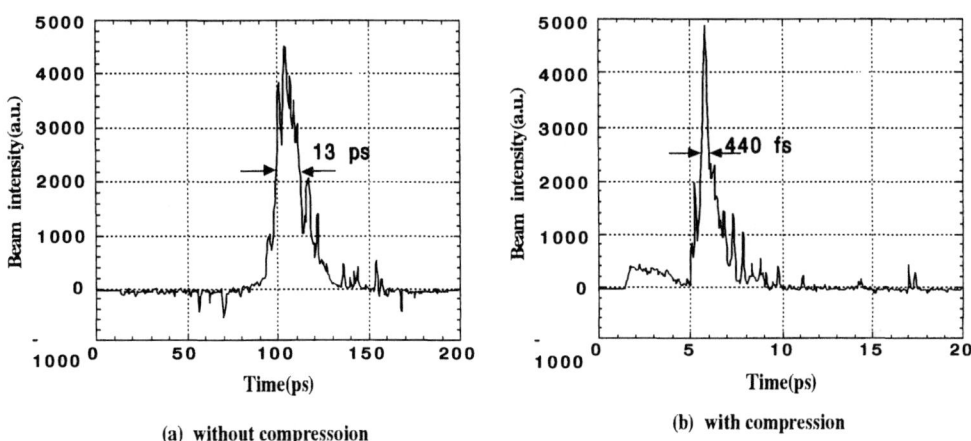

(a) without compressoion

(b) with compression

FIGURE3. Pulse shapes without and with compression measured by the femtosecond streak camera.

Hajima et al. analyzed the effect of normal space charge force, noninertial space charge force and coherent radiation force on the emittance, energy spread and bunch shape by their computer code and we carried out the measurement at this linac in 1998[4]. They calculated the total emittance growth of 29 pai mm.mrad, energy shift of 240keV and bunch broadening from 100fs to 500fs (FWHM) assuming 1 nC per bunch at the chicane. We measured the energy shift of 94keV, but not the emittance growth and bunch broadening for 0.6 nC per bunch. Calculated energy shift of 105keV for 0.6 nC agrees well with the measurement. Calculated emittance growth was 5 pai mm.mrad for 0.6 nC, but we could not measure it. Disagreement could be attributed to the change of the beam parameters during the measurement.

We added one NEG (Getter) vacuum pump to the gun last April so that the vacuum was improved to less than 1×10^{-9} torr even during RF feeding. Then QE was also improved from 4.5×10^{-5} to 1.5×10^{-4} and 7nC/bunch for 250μJ 267nm light (the 3rd harmonics of the Ti:Sapphire fundamental) was achieved at the exit of the gun [3]. However, the damage of the copper cathode was found from the viewport this February. The surface is uniformly rugged with tens micro-m craters so that it can attribute to RF discharge during the poor vacuum term. Re-measured emittance was 20 pai mm. mrad, which was rather degraded from our best value of 6 pai mm. mrad.

SUBPICOSECOND SYNCHORONIXATION BETWEEN ELECTRON BUNDH AND LASER

In order to achieve subpicosecond linac-laser synchronization, we have designed and constructed a new system introducing the most advanced technologies as shown in the left-hand-side of Fig.1. The design is based on the achievement at FELIX of FOM where the timing jitter between the electron and FEL pulse is 400fs (rms) [5]. The technologies are; Kerr-lens-mode-locked Ti:Sapphire laser with the timing stabilizer at the 9th harmonics RF (Coherent, Synchro-lock), the compact laser amplifiers, one stable 15 MW klystron RF power supplier (Mitsubishi, Modifed PV-3015), tempreture-controlled (within 1□)laser clean room (CLASS 10,000) and the vacuum laser transport line and so on. Timing jitters between the electron and laser pulses in the previous and new systems are evaluated in Table.1.

K. Kobayashi et al., achieved 77fs (rms) timing jitter at their mode-locked Ti:Sapphire laser [6]. Concerning the timing jitter at a laser oscillator, recent passive Ti:Sapphire passive mode-locker using the Kerr lens effect (for example, Coherent, Mira) is superior to conventional active Ti:Sapphire mode-lockers using an A/O crystal (for example, Spectra Physics, Tsunami) . The nonlinear Kerr lens effect in the Ti:Sapphire crystal enables much faster feedback tuning of the resonance frequency of the laser pulses. The timing stabilizer operating at the 9th harmonics of the input RF of 79.3 MHz is also effective. This is because the spectral noise due to the timing jitter is enhanced by the square of the harmonic number, while that due to the power fluctuation is constant [3]. When we used two

14

TABLE1. Measured and designed timing jitters (rms) in the previous and new systems.

	Previous system (measured)		New system (design)
	Thermionic	Photocathode	Photocathode
RF linac (σ_{rf})	a few ps	a few ps	300fs
Mutual jitter between two klystrons (σ_{rf2})	a few ps	a few ps	~0fs
Laser (σ_{laser})	< 3ps	< 3ps	100fs
Laser mode-locker	Active by A/O		Passive by Kerr lens
Timing stabilizer	at the fundamental (79.3MHz)		at 9th harmonics (713.7MHz)
Total	3.7ps	3.5ps	320fs

independent klystrons for the RF gun and accelerating tube, respectively, the electron bunch suffered from the mutual RF fluctuation between the two klystrons. In order to avoid this effect, we have chosen to use one klystron to feed RF into them. The electron beam jitter corresponds to the jitter of the electron bunch behind the chicane-type magnetic pulse compression due to fluctuations of RF power and phase, which are evaluated by the klystron performance data and PARMELA simulation. Here we used the RF voltage fluctuation of 0.02 dB (rms) and phase fluctuation of 0.2 deg (rms) based on the measurement using another same type of the klystron. The jitter of the time-of-flight gives the timing jitter of 300fs (rms) in the linac. The total timing jitter between the electron and laser pulses is expected to be ~320 fs (rms), according to eq. (1), in the new system,

$$\sigma_{total}{}^2 = \sigma_{rf}^2 + \sigma_{laser}^2 , \tag{1}$$

where $\sigma_{laser} = 100$fs is used.

We measured and evaluated current precision of electron-laser synchronization by the femtosecond streak camera (Hamamatsu FESCA-200), where the time resolution is 200fs, as shown in Fig.4. We operated the timing, stablizer at the fundamental (79.33MHz) and 9th harmonics (714.97 MHz) modes. The light-hand side shows the synchronized femtosecond electron and laser pulses. We stacked the same data more than 100 times, measured their time differences and drew the histogram for the two modes. We investigated that the slow drift attributed to the temperature change of the accelerating tube of the linac with not precise the old water cooling system ($\leq 1°C$). The jitter riding on the drift at 9th harmonics mode

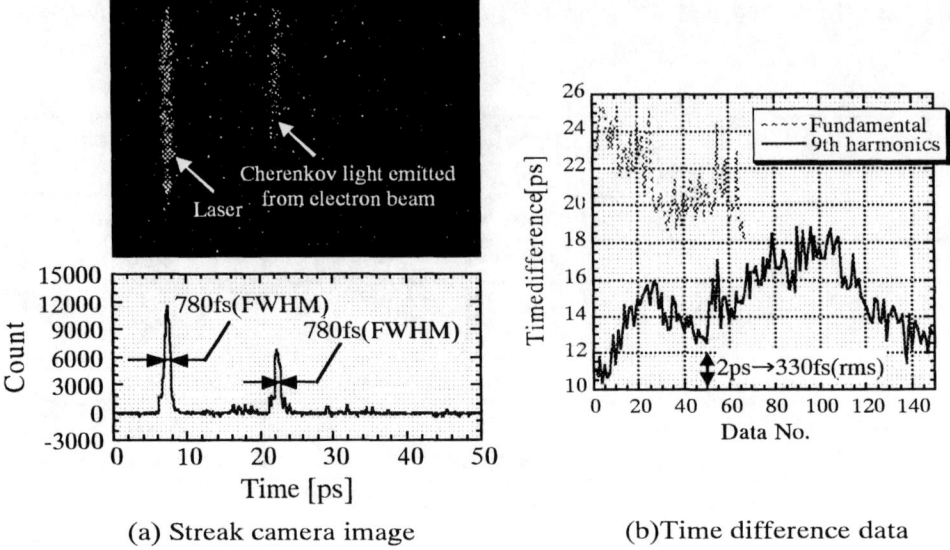

(a) Streak camera image   (b)Time difference data

FIGURE4. Measured results of the linac-laser synchronization.

is ~2ps (p-p), which corresponds to ~330fs (rms), for a few minutes. This is because one data set was obtained in about a minute. This is close to the design value of 320fs. It is also clearly observed that the jitter at the 9th is superior compared to that at the fundamental. We are going to introduce a more precise water cooling system within 0.01° so as to establish 330fs synchronization for longer time.

ELECTRON BUNCH DIAGNOSIS

For the purpose to diagnose ultrashort electron bunches, several methodolozies such as the zero-phasing method, the strip-line monitor, the femtosecond streak camera and the CTR interferometer have been developed and evaluated. We have been using the femtosecond streak camera since 1995 as a major diagnostic tool [8]. Since its limitation of resolution is 200fs (FWHM), we introduced the CTR Michelson interferometer [9,10,11,12] after having tested the Martin-Pupplett interferometer [13]. Further, we have tried the far-infrared polychromator, which consists of one grating, 10ch InSb bolometer array and cryostat, for single shot measurement [14]. Power spectrum of CTR can be obtained shot by shot by using it. The shot-by-shot diagnosis is very useful and important to investigate the fluctuation and instability of RF linacs. Thus, we have prepared the diagnostic tools in both temporal and spectral regions. It is very important to have more than two methodologies based on different physics to confirm the results.

We have simultaneously performed the comparison of measurement of subpicosecond electron pulses among the femtosecond streak camera, the Michelson interferometry and the polychromator measurement as shown in Fig.5. The measurement was done using the first linac with the 90keV thermionic gun. Since it can also generate a subpicosecond electron bunch [8], it is not a problem for checking their time-resolutions. We measured the transition radiation in the far-infrared region emitted by an electron bunch at the Al-foil in the air after the 50 μm-thick Ti window at the end of the first linac with the thermionic gun. We used liquid-He-cooled Si bolometers as a detector for the far-infrared radiation. The major beam parameters are as follows: the energy was 34MeV, the pulse length is about 600fs (FWHM) and the electron charge per bunch is controlled to be from 10 to 100 pC avoiding the over-scale of the detectors.

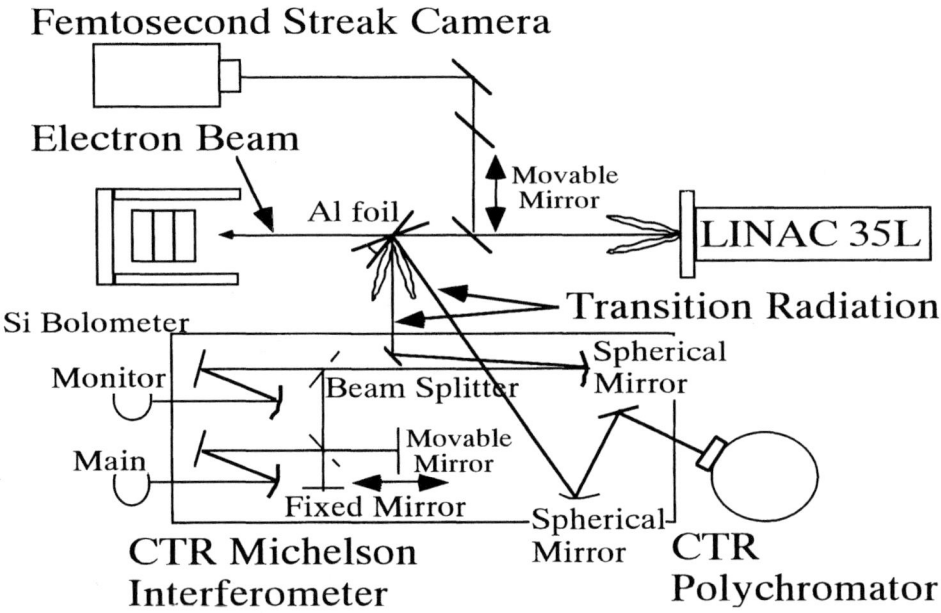

FIGURE 5. Electron bunch diagnostic setup.

The longitudinal distribution is evaluated based on the theory of coherent transition radiation. The longitudinal bunch form factors obtained by the two methods were rather limited because of the nonuniform transparency of the 100 μm-thick Mylar beam splitter in the Michelson interferometer and measurement region which depends on the grating pitch (1.0 mm) installed in the polychromator.

Therefore, we have to adopt theoretical extrapolation assuming the Gaussian or exponential distributions out of the range, referring to the pulse shape measured by the streak camera. The CTR spectrum calculated from the interferogram and by the polychromator are shown by the solid curves and the transparency of the beam splitter by the dashed curve in Fig.6. From the figure, we decided to use the experimental data in the range of 9.5 to 18.0 cm^{-1} for the analysis in the interferometry, while the measurable range of the polychromator was already determined from 12.2 to 26.2 cm^{-1} discretely by the 1mm grating pitch.

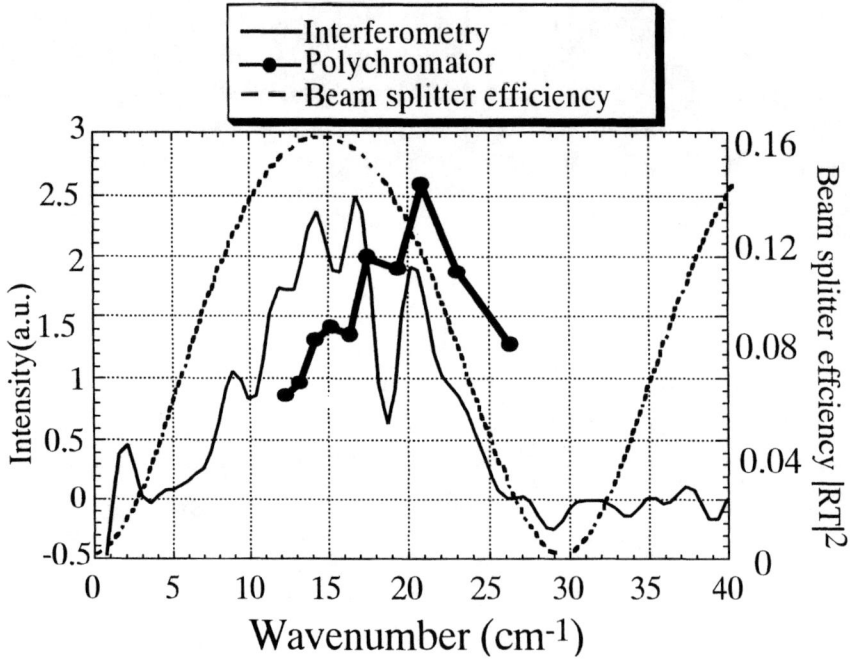

FIGURE 6. Spectrum of CTR.

Finally, we reconstructed the longitudinal bunch distributions after using Kramers-Kronig relation to derive the phase information. The result of the subpicosecond pulse measurement by the interferometry and that by the polychromator were 650 fs ps at FWHM as shown in Fig.7. Typical result by the streak camera is also shown in the same figure. Here we have got reasonable agreement and confirm the enough reliability of the diagnostics methods by the CTR measurement.

Characteristics of the three methods are summarized in Table.2. Up to 200fs (FWHM), the streak camera is the best because direct bunch shape can be obtained shot by shot. However, we expect that the CTR methods are promising for the shorter electron beam than 200fs with better resolution because the coherent spectrum shifts from the far-infrared region to the infrared or visible region where the sensitivity of the detector becomes better. Especially, the polychromator can be expected to the most powerful method because of the advantage of diagnostics by a single shot.

Recently, the fluctuation method [15] to evaluate subpicosecond bunch length from the fluctuation of both intensity and spectrum of incoherent Cherenkov radiation is under way.

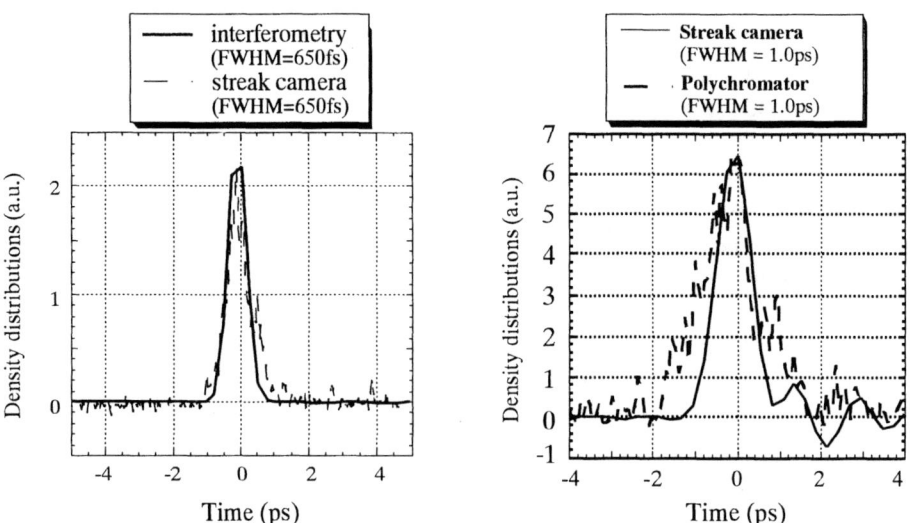

FIGURE7. Bunch distributions obtained by the femtosecond streak camera, CTR Michelson interferometer and far-infrared polychrometer.

TABLE 2. Characteristics of short electron bunch diagnostic methods.

	Measurement Source	Measurement Limitation (Reported)	Single-shot	In Vacuum	Non-destructive
Femtosecond Streak Camera	Cherenkov Radiation	200 fs (~240 fs)	O	O (Solid Radiator) X (Gas Radiator)	X
Coherent Radiation Interferometry	CTR(Coherent Transition Radiation) CDR(Coherent Diffraction Raditation)	Unlimitted (~120 fs)	X	O	O (CDR) X (CTR)
10ch Polychromator			O		
Fluctuation Method	Incoherent Radiations	Unlimitted (~ps)	O	O	O (DR) X (TR)
Zero-phasing Method	Energy Spectrum of Electron Pulse	Unlimitted (~100 fs)	X	O	X
S M A Monitor	Induced Voltage from Electron Pulse	~ps (~ps)	O	O	O

19

NUMERICAL AND EXPERIMENTAL WORK
ON THE PLASMA GUN

Many leading researches on plasma wakefield acceleration have continued here in this decade under the collaboration with KEK and JAERI [16]. Recently we proceed to the laser plasma linac to generate 10fs relativistic electron single bunch by a single 12TW 50fs laser pulse [17]. As the laser is more intense in a gas jet, the plasma wakefield changes from linear (sinusoidal) to nonlinear (sharper wavefront) and the election motion in plasma becomes from nonrelativistic to relativistic. Finally, beyond the critical value, the wake wavebreaking occurs so that the wave energy is transferred to longitudinal electron momentum. Those electrons are accelerated and bunched by the wakefield. The configuration and numerical result by the PIC (Particle In Cell) −2D code are shown in Figs.8 and 9, respectively. 25MeV(max), 12fs(FWHM) and ~2.8 pai mm.mrad(rms) electron single bunch with 10^{11} electrons is generated as shown in Fig.4. We plan to verify this idea and generation of ～10fs ultrashort electron bunch experimentally in near future. After we have measured the electron bunch, we plan to construct tens femtosecond time-resolved pump-and-probe analysis system using a laser beam splitter and optical delay line (see Fig.9). Since the synchronization is done fully passively, the positioning precision of 5μm at the delay line corresponds to 33fs time delay without the timing jitter.

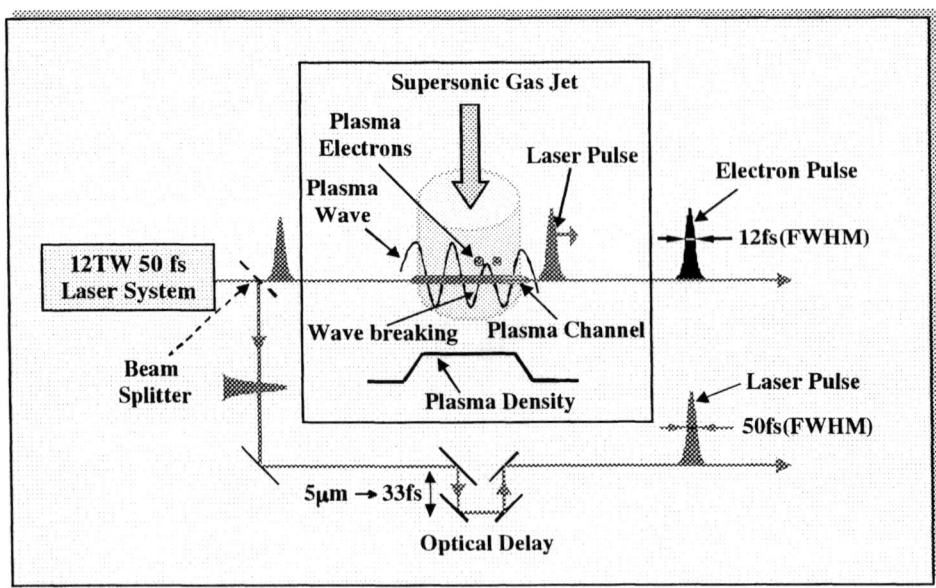

FIGURE 8. Configuration of research and development for the plasma gun.

In the experiment, an OAP (off-axis parabolic) mirror focused 50 fs laser pulses with peak power up to 12TW near the top of a supersonic He gas jet (pressure of >1000 psi). The calculated spot size and intensity of the laser above the jet were about 12.6 μm and 4×10^{18} W/cm^2 respectively. The charges per pulse and transverse profiles of the accelerated electrons were detected in the forward direction by using faraday cup and imaging plate, respectively. The charge/pulse was larger than 1nC and the transverse beam profile were near gaussian with about 7^0 divergence angle (FWHM) from the electron source (gas jet). Simulations in the near experimental regime indicate a simple

characteristic dependence of the beam energy on the angle of deflection via the relation $\cos\theta = (\gamma - 1/\gamma + 1)^{1/2}$[18], so that for an angle of 7^0 divergence, it is found that electrons with energy around 68 MeV were generated in well-collimated beam.

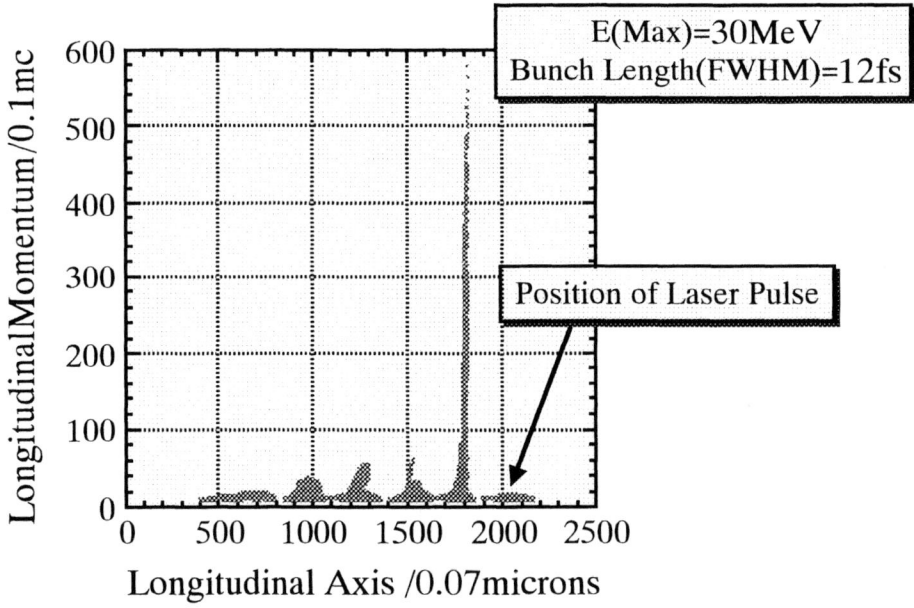

FIGURE 9. Numerical result of 12fs relativistic electron beam generation by PIC-2D.

CONCLUSION

Femtosecond S-band linac with the BNL/GUN IV RF gun photoinjector has recorded 240fs(FWHM), 6paimm. mrad(rms), 350pC/bunch at Nuclear Engineering Research Laboratory, University of Tokyo. Maximum charge per bunch at the exit of the photoinjector and QE were 7nC and 1.5×10^{-4}. After the cathode damage was found, re-measured emittance was 20 pai mm. mrad. Timing jitter between the femtosecond electron bunch and the drive laser for the injector was 330fs for the minutes and 1.9ps for hours. Numerical analysis on the plasma gun by the 12TW50fs Ti:Sapphire laser predicts to produce 10fs, 20MeV electron single bunch without any timing jitter from the drive multi-TW laser. Experimental verification is under way. We believe that the describes achievement and data give important mile stones for development of future SASE-FEL including LCLS.

REFERENCE

[1] X.J. Wang et al., *Nucl. Instrum. Meth. A.* **375**, 1996, pp.82-86.
[2] M.Uesaka et al., "Experimental Verification of Laser Photocathode RF Gun as an Injector for a Laser Plasma Accelerator", *Trans. Plasma Science*, 2000, in press.
[3] T. Kobayashi et al., Proc. of PAC2000, 2000, pp.1663-1665.
[4] R. Hajima et al., *Nucl. Instrum. Meth. A.* **429**, 1999, pp.264-268.
[5] G.M.H. Knippels et al., Optics Lett., **23**(22), 1998, pp.1754-1756,.
[6] K. Kobayashi et ak., *SPIE*, **3616**, 1999, pp.156-164.
[7] D. von der Linde, *Appl. Phys.* B39, 1986, pp.201-217.
[8] M. Uesaka et al., *Nucl. Instrum. Meth. A*, **406**, 1998, pp.371-379.
[9] Y. Shibata et al., *Phys. Rev. E*, **50**(2-B), 1994, pp.1479-1484.
[10] R. Lai et al., *Phys. Rev. E*, **50**(6), 1994, pp.R4294-R4297.
[11] H. Lihnet al., *Phys. Rev. E*, **53**(6), 1996, pp.6413-6418.
[12] T. Watanabe et al., *Nucl. Instrum. Meth. A*, **437**, 1999, pp.1-11.
[13] M. Uesaka et al., *Nucl. Instrum. Meth. A.* **410**(3), 1998, pp.424-430.
[14] J. Sugahara et al., *Proc. of the 1999 Particle Accelerator Conference*, pp.2187-2189, 1999.
[15] P. Catravas et al., *Phys. Rev. Lett.*, **82**(62), 1999, pp.5261-5264.
[16] H. Dewa et al., *Nucl. Instrum. Meth. A.*, **410**(3), 1998, pp.357-363.
[17] N. Hafz et al., *Nucl. Instrum. Meth. A*, **455**, 2000, pp.148-154.
[18] P. Mora et al., *Phys.Rev. E*, **53**, p.R2068, (1996).

Beam Collimation for LCLS Beam Emittance and Charge Control

C. B. Schroeder*, H.-D. Nuhn† and C. Pellegrini*

*Department of Physics and Astronomy, University of California, Los Angeles, CA, 90095, USA
†Stanford Linear Accelerator Center, Stanford University, Stanford, CA, 94309, USA

Abstract. In this paper we describe an electron beam collimator for the Linac Coherent Light Source. The collimator reduces both the transverse emittance and the total charge of the electron beam. Beam optics are designed to increase the beam spot size for collimation. The impedance of the collimator is calculated, and the wakefield induced emittance growth is determined.

INTRODUCTION

The Linac Coherent Light Source (LCLS) [1] is a proposed single-pass free-electron laser (FEL) which produces Ångstrom wavelength radiation using a multi-GeV electron beam produced at the Stanford Linear Accelerator Center (SLAC). The LCLS is designed such that the FEL process reaches saturation. The saturation length for present design parameters is 94 m, or about 16 power gain lengths. The power gain length is minimized, and therefore the gain increased for a given undulator length, when the transverse emittance of the electron beam matches the phase-space characteristics of the emitted radiation [2, 3, 4]

$$\varepsilon_n < \gamma \frac{\lambda}{4\pi} \, , \tag{1}$$

where ε_n is the normalized transverse emittance of the electron beam, λ is the radiation wavelength, and γ is the Lorentz factor of the electron beam. For radiation wavelengths in the Ångstrom regime, this condition is difficult to satisfy using conventional electron beam sources. The LCLS design parameters do not satisfy Eq. (1), and as a result, the power gain length is about a factor of two larger than what is predicted by one-dimensional (1D) FEL theory.

In this paper we propose to eliminate a portion of the four-dimensional (4D) transverse phase-space volume occupied by the electron beam through the use of two collimators. This will reduce both the transverse emittance and total charge of the electron beam. Reduction of the transverse emittance will allow the electron beam to satisfy Eq. (1), which will reduce the power gain length of the FEL process, and therefore, the total length of the undulator required to reach saturation. Reduction of the electron beam charge is also advantageous. Reduction of the electron beam charge will reduce excitation of wakefields throughout the beamline, lessening the degradation of beam quality by wakefield induced instabilities. The peak power of the FEL radiation will also be

CP581, *Physics of, and Science with, the X-Ray Free-Electron Laser*, edited by S. Chattopadhyay et al.
© 2001 American Institute of Physics 0-7354-0022-9/01/$18.00

reduced for smaller beam charge, thereby decreasing the damage to conventional x-ray focusing and transport optics after the undulator.

TRANSVERSE PHASE-SPACE REDUCTION

In this section we consider the effect of the collimation on the charge and transverse emittance of the electron beam. We assume that the initial transverse 4D phase-space distribution of the electron beam is Gaussian and has the form

$$f_\perp(\vec{r},\vec{r}') = \frac{1}{4\pi^2\sigma^4 k_\beta^2} \exp\left[-\frac{(k_\beta^2 r^2 + r'^2)}{2\sigma^2 k_\beta^2} \right], \tag{2}$$

where $\vec{r} = (x,y)$ and $\vec{r}' = (dx/dz, dy/dz)$. Here we are assuming a symmetric matched beam such that the rms spot size is $\sigma = \sigma_x = \sigma_y$ and the normalized rms transverse emittance is $\varepsilon_n = \gamma\varepsilon_x = \gamma\varepsilon_y = \gamma k_\beta \sigma^2$.

The collimation procedure is to pass the beam through a (round) collimator, limiting the extent of the beam in two dimensions of the 4D phase-space. The beam is then allowed to phase-advance 90 degrees in both transverse directions. After the phase-advance, the beam is passed through a second (round) collimator, which will limit the extent of the electron beam in the remaining two dimensions of the 4D transverse phase-space. The beta-functions of the beam are assumed to be equal at the positions of the two collimators. After passing through the two collimators, the transverse phase-space distribution is a cut-off Gaussian such that the final number of electrons N_f is given by

$$\frac{N_f}{N_i} = \int_0^{b_1} (2\pi r dr) \int_0^{k_\beta b_2} (2\pi r' dr') f_\perp(r,r') = \prod_{j=1}^{2} \left(1 - e^{-\frac{b_j^2}{2\sigma^2}} \right), \tag{3}$$

where N_i is the initial number of electrons, b_1 is the radius of the first collimator, and b_2 is the radius of the second collimator. The transverse rms emittance will be reduced such that,

$$\frac{\varepsilon_f^2}{\varepsilon_i^2} = \prod_{j=1}^{2} \left[1 - \exp\left(-\frac{b_j^2}{2\sigma^2} \right) \right] \left[1 - \left(1 + \frac{b_j^2}{2\sigma^2} \right) \exp\left(-\frac{b_j^2}{2\sigma^2} \right) \right], \tag{4}$$

where $\varepsilon_i = k_\beta \sigma^2$ is the initial transverse rms emittance and ε_f is the final transverse rms emittance after collimation. Figure 1 shows N_f/N_i and $\varepsilon_f/\varepsilon_i$ versus collimator radius normalized to the initial rms spot size b/σ for two collimators of equal radii $b = b_1 = b_2$. As the figure shows, 90% of the beam can be removed by choosing $b \approx \sigma$. The figure also implies that the efficiency of the FEL process, which is determined by the FEL parameter ρ, will always increase as a result of the collimation, i.e., $\rho_f/\rho_i = (N_f\varepsilon_i/N_i\varepsilon_f)^{1/3} > 1$.

For LCLS parameters ($\gamma = 28077$, $\lambda = 1.5$ Å, and $\varepsilon_i = 1.1$ mm mrad), a matched beam Eq. (1) requires $\varepsilon_f/\varepsilon_i \simeq \gamma\lambda/4\pi\varepsilon_i \simeq 0.27$. This can be achieved with collimators of radii $b/\sigma \simeq 1.6$, which reduces the charge to $N_f/N_i \simeq 0.53$.

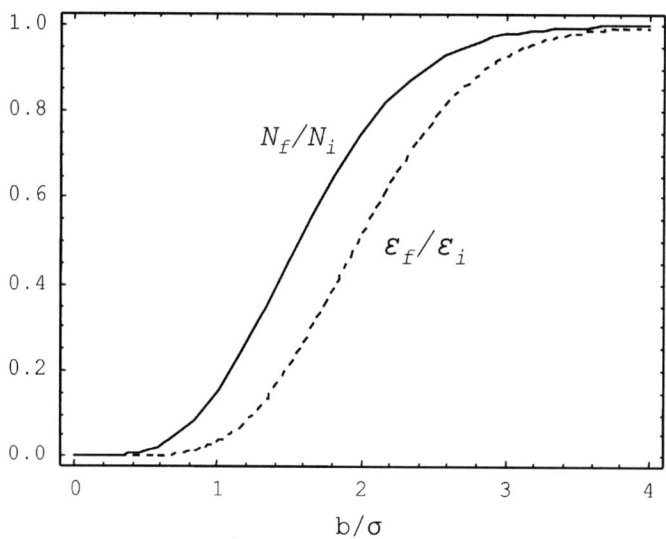

FIGURE 1. Ratio of rms transverse emittance after collimation to before collimation $\varepsilon_f/\varepsilon_i$ (dashed line) and ratio of number of electrons in the electron bunch after collimation to before collimation N_f/N_i (solid line) versus collimator radius normalized to the initial rms spot size b/σ for two collimators of equal radii $b = b_1 = b_2$.

COLLIMATOR BEAM LINE

We consider placement of the collimator after the first linac (L0) following the photo-cathode gun on the LCLS beam line. The electron beam parameters after the L0 linac are listed in Table 1 [1]. This choice of location restricts the total length of the collimator design (optics) to less than the distance between linac L0 and linac L1, which is approximately 14 m.

The initial rms spot size of the beam after L0 is $\sigma_x = \sqrt{\beta\varepsilon_n/\gamma} = 193$ μm. This small rms spot size and collimator fabrication limitations, requires the use of optics to increase the rms spot size (i.e., the beta function) before collimation. The beam line designed for this purpose uses 19 quadrupole magnets each of length 10 cm. Quadrupole strengths vary from $k_q = -57.95$ m^{-2} to $k_q = 50.78$ m^{-2}, where the quadrupole strength is related

TABLE 1. Electron beam parameters at end of L0 linac

beam energy	150.5 MeV
beam charge	1.0 nC
rms beam duration σ_t	2.3 ps
normalized rms transverse emittance	1.1 πmm-mrad
rms relative energy spread	0.13%
beta function β	10 m

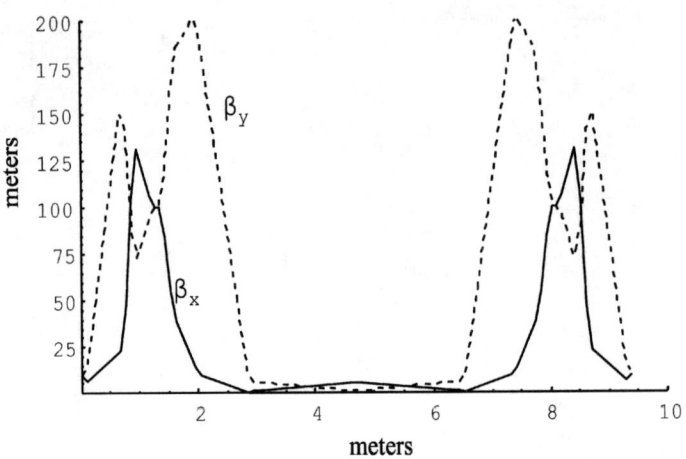

FIGURE 2. Beta functions β_x (solid line) and β_y (dashed line) versus distance along the beam line.

to the magnetic field gradient by $k_q[m^{-2}] = 0.2998\ g[T/m]/p[GeV/c]$. The length of the collimators are assumed to be < 6 cm. The total length of the optics is $L_{total} = 9.393$ m. The quadrupole strengths and separations were calculated to achieve beam twiss parameters at specified locations along the beam line using the MAD program version 8.1 [5].

Figure 2 shows the beta functions for both transverse directions versus propagation distance along the beam line. Four quadrupoles are used to increase the beta functions by a factor of 10 from their initial value. The first collimator is placed at $z = 1.312$ m, where the beam twiss parameters are $\beta_x = \beta_y = 100$ m and $\alpha_x = \alpha_y = 0$. Four additional quadrupoles are used to reduce the beta functions to $\beta_x = (1 + \sqrt{2})/(1 - \sqrt{2})$ m and $\beta_y = 1$ m with $\alpha_x = \alpha_y = 0$ at $z = 2.956$ m. These small beta functions are necessary to allow for a large phase-advance ($\Delta\mu = \pi/2$) over a relatively short distance (< 10 m). The electron beam then passes through a quarter-wave transformer [6], which provides the 90 degree phase-advance. Figure 3 shows the phase-advance for each transverse direction. Note the $\pi/2$ phase-advance in each transverse direction as the beam propagates from $z = 2.956$ m to $z = 6.437$ m. Four quadrupoles are used to return the twiss parameters to $\beta_x = \beta_y = 100$ m and $\alpha_x = \alpha_y = 0$ at $z = 8.081$ m where the second collimator is placed. Four additional quadrupoles are used to return the beam to the initial conditions $\beta_x = \beta_y = 10$ m and $\alpha_x = \alpha_y = 0$ at the end of the collimator beam line $z = 9.393$ m.

Although it is perhaps more elegant to construct a two lens magnification system (where each lens would be a symmetric quadruplet quadrupole array to provide equal focusing in both transverse planes) to increase the beta functions, constraints on the length of the beam line and maximum quadrupole strength due not permit such a solution for this application.

Figure 4 shows the rms spot sizes in both transverse directions over the length of the beam line. With $\beta_x = \beta_y = 100$ m, the electron beam rms spot size is $\sigma_x = \sigma_y = 611\ \mu$m.

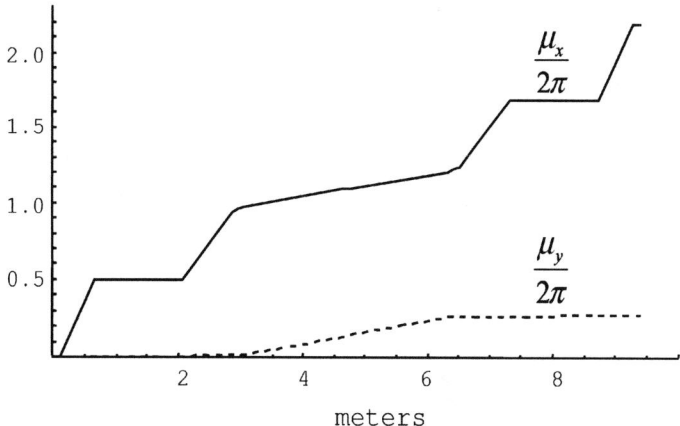

FIGURE 3. Phase-advance μ_x (solid line) and μ_y (dashed line) versus distance along the beam line.

Reduction of the beam charge such that $N_f/N_i = 0.53$ requires that the collimators to have radii of $b = 987\ \mu m$.

COLLIMATOR IMPEDANCE

Transverse emittance growth owing to wakefields excited by the passage of the beam through the collimators could set a bound on the transverse emittance reduction achievable by collimation; although previous experimental work [7] has achieved 90% removal of electron beam charge through collimation and has not reported wakefield induced transverse emittance growth. We consider emittance growth from geometric (i.e., resulting from the changing beam pipe shape) and resistive-wall wakefields.

Diffraction Model

The high-frequency behavior of the impedance can be approximated using diffraction theory [8, 9]. Here we are considering wakefield frequencies which satisfy $1 < \omega b/c \ll \gamma$. Using Kirchhoff diffraction theory [10], the axial field produced by the beam interacting with the structure (i.e., the diffracted field) is given by performing the integral over the surface S of the collimator

$$E_z = \hat{z} \cdot \hat{n} \int_S dS \frac{\partial G}{\partial r} \hat{r} \cdot \vec{E}_b,$$ (5)

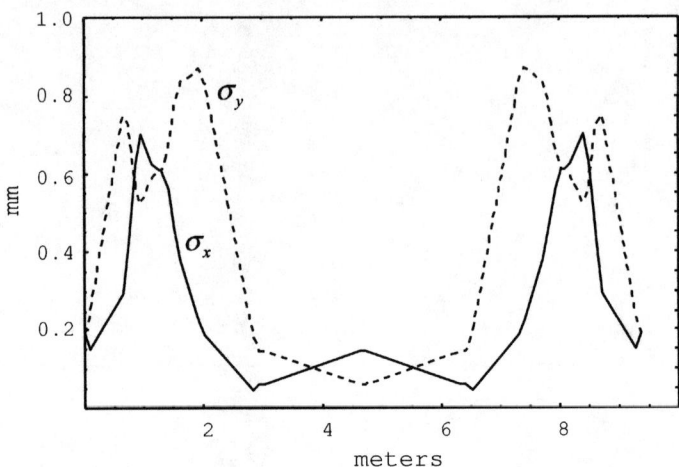

FIGURE 4. Electron beam rms transverse spot size σ_x (solid line) and σ_y (dashed line) versus distance along the beam line.

where $\vec{E}_b$ is the electric field of the beam in a straight pipe (i.e., the source field)

$$\hat{r} \cdot \vec{E}_b = \frac{2q}{cr} e^{ikz}, \tag{6}$$

and G is the Green function of the wave equation

$$\left(\nabla^2 + k^2\right) G(r, r') = -\delta(r - r'). \tag{7}$$

The Green function can be written as [10]

$$G = \frac{i}{8\pi} \int_{-\infty}^{\infty} dl e^{il(z-z')} \left[J_0(r_<s) H_0^{(1)}(r_>s) + 2 \sum_{m=1}^{\infty} J_m(r_<s) H_m^{(1)}(r_>s) \cos m\phi \right], \tag{8}$$

where J_m is the mth-order Bessel function, $H_m^{(1)}$ is the mth-order Hankel function of the first kind, $s = (k^2 - l^2)^{1/2}$, and $(r_>, r_<)$ is the greater and lesser of r and r'.

We will consider the geometry of a beam interacting with a cylindrical collimator of length L and radius b placed between beam pipe of radius a (with $a > b$). This geometry can be considered a zeroth-order collimator design. With this geometry, using Eqs. (5), (6), and (8), the longitudinal impedance is

$$
\begin{aligned}
Z_\parallel^{(0)} &= \frac{1}{q} \int dz e^{-ikz} E_z \\
&= \frac{1}{q} \int dz e^{-ikz} (\hat{z} \cdot \hat{n}) \int_S dS \frac{\partial G}{\partial r} \left(\frac{2q}{cr} e^{ikz} \right) \\
&= \frac{i}{2c} \int_{-\infty}^{\infty} dl \int_{-\infty}^{\infty} dz e^{i(l-k)z} \left(1 - e^{i(l-k)L} \right) J_0(rs) \left[H_0^{(1)}(as) - H_0^{(1)}(bs) \right]
\end{aligned}
$$

28

$$= \frac{2}{c} \int_{-\infty}^{\infty} \frac{dl}{l-k} J_0(bs) \left[H_0^{(1)}(bs) - H_0^{(1)}(as) \right] \sin^2[(l-k)L/2]. \tag{9}$$

The real (resistive) part of the longitudinal impedance is

$$\Re Z_\parallel^{(0)} = \frac{Z_o}{2\pi} \int_{-k}^{k} \frac{dl}{l-k} J_0(bs) \left[J_0(bs) - J_0(as) \right] \sin^2[(l-k)L/2], \tag{10}$$

where $Z_o = 4\pi/c$ is the impedance of free space. Note that, it follows from the Cauchy theorem that the imaginary part of the impedance can be determined from the real part through application of a Hilbert transform. The integral in Eq. (10) can be evaluated numerically. In the limit $kL \gg ka > kb > 1$, Eq. (10) can be approximated as

$$\Re Z_\parallel^{(0)} \simeq \frac{Z_0}{2\pi} \ln \left(\frac{a}{b} \right). \tag{11}$$

The diffraction model can be extended to provide an estimation of the transverse impedance [11]. The real (resistive) part of the monopole mode longitudinal impedance Eq. (11) can be expressed as the ratio of diffracted power $P_D^{(0)}$ to the square of the beam current

$$\Re Z_\parallel^{(0)} = \frac{P_D^{(0)}}{I_0(\omega)^2}, \tag{12}$$

where the monopole mode current I_0 produces the radial electric field at the collimator radius $E = Z_0 I_0 / 2\pi b$. Consider the mth moment current of strength I_m. The real part of the mth mode longitudinal impedance can be expressed as the ratio of the diffracted power from this mode to the square of the current strength

$$\Re Z_\parallel^{(m)} = \frac{P_D^{(m)}}{I_m(\omega)^2}. \tag{13}$$

The radial electric field produced by the mth moment of the beam current at the collimator radius is $E_m = (Z_0/\pi b^{m+1}) I_m \cos(m\theta)$, and therefore the ratio of the total power diffracted from the mth mode around the perimeter of the collimator to the diffracted monopole power is

$$\frac{P_D^{(m)}}{P_D^{(0)}} = \frac{2}{b^{2m}} \left(\frac{I_m}{I_0} \right)^2. \tag{14}$$

It follows from Eqs. (12), (13), and (14) that the mth mode (for $m > 0$) longitudinal impedance is

$$\Re Z_\parallel^{(m)} = \frac{2}{b^{2m}} \Re Z_\parallel^{(0)}. \tag{15}$$

Using the Panofsky-Wenzel theorem and Eqs. (11) and (15), the real part of the transverse m^{th} moment impedance is

$$\Re Z_\perp^{(m)} = \frac{c}{\omega} \Re Z_\parallel^{(m)} = \frac{2c}{b^{2m}\omega} \Re Z_\parallel^{(0)} = \frac{Z_0 c}{\pi b^{2m}\omega} \ln \left(\frac{a}{b} \right). \tag{16}$$

Emittance Growth

The wakefields produced by the changing beam pipe can produce growth in the emittance of the electron beam. The wakefields can be calculated by computing the inverse Fourier transform of the transverse impedance

$$W_m = \frac{2}{\pi} \int_0^\infty d\omega \sin(\omega z/c) \Re Z_\perp^{(m)}(\omega) = \frac{4}{b^{2m}} \ln(a/b). \tag{17}$$

The transverse momentum kick to a test charge from these wakefields is

$$c\Delta\vec{p}_\perp = -e \sum_m I_m W_m m r^{m-1} \left(\hat{r}\cos(m\theta) - \hat{\theta}\sin(m\theta) \right) \tag{18}$$

$$= -eq \left\{ W_1 \left[\langle x \rangle \hat{x} + \langle y \rangle \hat{y} \right] + 2W_2 \left[Q_n (x\hat{x} - y\hat{y}) + Q_s (y\hat{x} + x\hat{y}) \right] + \right.$$

$$\left. + 3W_3 \left[S_n \left([x^2 - y^2]\hat{x} - 2xy\hat{y} \right) + S_s \left(2xy\hat{x} + [x^2 - y^2]\hat{y} \right) \right] + \dots \right\},$$

where $Q_n = \langle x^2 - y^2 \rangle$, $Q_s = \langle 2xy \rangle$, $S_n = \langle x^3 - 3xy^2 \rangle$ and $S_s = \langle 3x^2 y - y^3 \rangle$ are the normal and skew quadrupole and sextupole moments of the beam distribution in Cartesian coordinates respectively.

The normalized rms transverse emittance of the beam after passing through the collimator ε_f due to the momentum kick $\hat{x} \cdot \Delta\vec{p}_\perp$ [given by Eq. (18)] will be

$$\varepsilon_f^2 = \sigma_x^2 \left[\langle (p_x + \Delta p_x)^2 \rangle - \langle p_x + \Delta p_x \rangle^2 \right] - \left[\langle x(p_x + \Delta p_x) \rangle - \langle x \rangle \langle p_x + \Delta p_x \rangle \right]^2$$

$$= \varepsilon^2 + \sigma_x^2 \left(2\langle p_x \Delta p_x \rangle - 2\langle p_x \rangle \langle \Delta p_x \rangle + \langle \Delta p_x^2 \rangle - \langle \Delta p_x \rangle^2 \right)$$

$$- 2\sigma_{xp_x} \left(\langle x\Delta p_x \rangle - \langle x \rangle \langle \Delta p_x \rangle \right) - \left(\langle x\Delta p_x \rangle - \langle x \rangle \langle \Delta p_x \rangle \right)^2, \tag{19}$$

where ε^2 is the unperturbed emittance, $\sigma_x^2 = \langle x^2 \rangle - \langle x \rangle^2$ is the rms beam spot size and $\sigma_{xp_x} = \langle xp_x \rangle - \langle x \rangle \langle p_x \rangle$ is the linear correlation between position and momentum. Equations (18) and (19) show that the dipole and quadrupole moments of the beam distribution will not contribute to the rms emittance growth. The dipole moment will cause a uniform kick to the beam and a shift in the beam centroid, which may cause beam breakup. The quadrupole moment will provide a linear force on the beam and tend to cause the rms beam size of the bunch tail to grow. The leading-order contribution to the emittance growth will be from the sextupole moment of the beam distribution.

If there are initially no correlations between transverse position and momentum [i.e., $\langle p_x f(x) \rangle = 0$, where $f(x)$ is any function of x] and no initial transverse drift $\langle p_x \rangle = 0$, then Eq. (19) can be written as

$$\varepsilon_f^2 = \varepsilon^2 + \sigma_x^2 \left(\langle \Delta p_x^2 \rangle - \langle \Delta p_x \rangle^2 \right) - \left(\langle x\Delta p_x \rangle - \langle x \rangle \langle \Delta p_x \rangle \right)^2. \tag{20}$$

The change in emittance can be evaluated using Eqs. (18) and (20) for an arbitrary beam distribution. For example, consider a displaced cut-off Gaussian distribution at the exit

30

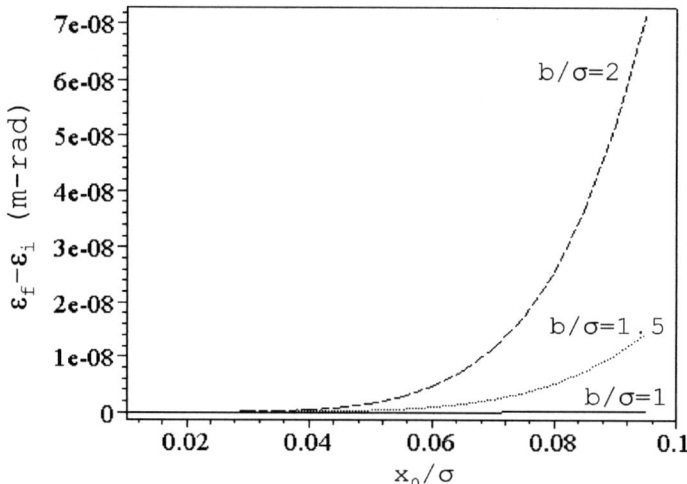

FIGURE 5. Emittance growth (m-rad) due to sextupole geometric wakefield versus initial transverse displacement of a Gaussian beam centroid from the axis normalized to the initial rms beam radius x_o/σ, for several collimator radii normalized to the initial rms beam radius $b/\sigma = 1$, 1.5, and 2.

of the collimator such that

$$
f_\perp(x,y) = \begin{cases} \frac{1}{2\pi\sigma^2} e^{-[(x-x_o)^2 + y^2]/2\sigma^2} & \text{for} \quad \sqrt{x^2 + y^2} < b \\ 0 & \text{for} \quad \sqrt{x^2 + y^2} \geq b \end{cases} , \tag{21}
$$

where x_o is the initial off-axis beam centroid displacement entering the collimator. Figure 5 shows the emittance growth due to the sextupole wakefield using this distribution versus initial beam offset for a given collimator radius. Here we are using the beam parameters in Table 1 with the collimator parameter $a/b = 100$. The figure shows that the emittance growth is less than 1% for an error in beam alignment of less than 5% . The figure also shows a reduction of emittance growth for small collimator radii owing to the reduction in the sextupole moment for smaller radii.

Resistive-Wall Wakefields

The resistive-wall wakefields, due to the finite conductivity of the walls, will also produce a transverse momentum kick to the electron beam. The resistive-wall wakefields for $m > 0$ are [12]

$$
W_m^{(\text{res})} = -\frac{2L}{\pi b^{2m+1} \sqrt{\sigma_c \sigma_t}} , \tag{22}
$$

31

where σ_c is the conductivity of the collimator walls. Comparing Eqs. (17) and (22), one can see that the ratio of resistive-wall to high-frequency geometric wakefields is

$$\left| \frac{W_m^{(\text{res})}}{W_m^{(\text{geo})}} \right| = \frac{L}{2\pi b \ln(a/b) \sqrt{\sigma_c \sigma_t}} \ll 1 \qquad (23)$$

for the parameters considered in this paper. For the beam parameters in Table 1 with collimator parameters $b = 500 \ \mu$m, $L = 3$ cm, $a/b = 100$, and $\sigma_c = 5 \times 10^{17}$ s^{-1}, this ratio is $\sim 10^{-3}$. Therefore the geometric wakefields will dominate the resistive-wall wakefields in this parameter regime.

SUMMARY

In this paper we have examined the use of a collimator to reduce the transverse phase-space and the total charge of the electron beam proposed for the LCLS. This will allow the transverse emittance of the beam to match the transverse phase-space of the radiation, thereby reducing the power gain length and the total length of the undulator required to achieve saturation of the FEL process. Reduction in the charge of the beam will also reduce wakefield-induced instabilities along the beam line. The impedance of the collimator was calculated, and the wakefield induced emittance growth due to the beam passing through the collimator was shown to be small for the parameter regime under consideration.

ACKNOWLEDGMENTS

This work was supported be the U. S. Department of Energy, under Contract No. DE-FG03-92ER40693.

REFERENCES

1. Linac Coherent Light Source (LCLS) Design Study Report, Tech. Rep. SLAC-R-521, Stanford Linear Accelerator Center (1998).
2. Xie, M., *Nucl. Instrum. Methods*, A445, 59 (2000).
3. Yu, L. H., Krinsky, S., and Gluckstern, R. L., *Phys. Rev. Lett.*, 64, 3011 (1990).
4. Kim, K.-J., *Phys. Rev. Lett.*, 57, 1871 (1986).
5. Grote, H., and Iselin, F. C., The MAD (Methodical Accelerator Design) Version 8.1 User Reference Manual, Tech. rep., CERN (1990).
6. Brown, K. L., and Servranckx, R. V., First- and Second-Order Charged Particle Optics, Tech. Rep. SLAC-PUB-3381, Stanford Linear Accelerator Center (1984).
7. Qiu, *et al..*, X., *Phys. Rev. Lett.*, 76, 3723 (1996).
8. Gluckstern, R. L., *Phys. Rev. D*, 39, 2773 (1989).
9. Heifets, S. A., and Kheifets, S. A., *Phys. Rev. D*, 39, 960 (1989).
10. Jackson, J. D., *Classical Electrodynamics*, Wiley, 1975.
11. Bane, K., and Sands, M., *Part. Accel.*, 25 (1990).
12. Chao, A., *Physics of Collective Beam Instabilities in High Energy Accelerators*, Wiley, 1993.

Wake Fields Effects due to Surface Roughness in a Circular Pipe

M. Angelici*, F. Frezza*, A. Mostacci*[†] and L. Palumbo*

*Università di Roma "La Sapienza", Roma, Italy
[†]CERN, Geneva, Switzerland

Abstract. The problem of the wake field generated by a relativistic particle travelling in a long beam pipe with rough surface has been revisited by means of a standard theory based on the hybrid modes excited in a periodically corrugated waveguide with circular cross section. Slow waves synchronous with the particle can be excited in the structure, producing wake fields whose frequency and amplitude depend on the depth of the corrugation.

INTRODUCTION

The effect of surface roughness is a subject arisen in the design of machines with extremely short bunches of the order of tens of microns. In this case, in fact, the surface roughness may be a source of wake fields which might significantly increase the beam emittance and the energy spread. In the LCLS (Linac Coherent Light Source) the surface roughness is due to residual defects in workmanship, and may be responsible of the longitudinal emittance growth due to wakefields.

In this paper we review the problem of the wake fields produced by an ultra-relativistic charge traveling inside a beam tube with a periodic corrugation making use of a standard theory based on the hybrid modes propagating in the waveguide [1, 2]. The paper is structured as follows: in the First Section we describe the method used, in the Second Section we present the results: first the dispersion relation for the fields and the frequency where the wave is synchronous with the charge can be excited; then, the amplitude of the field excited by the charge, the wake function and the coupling impedance; the results are applied to the LCLS case. In the Appendix are explained some calculation.

THE METHOD

Let us consider a periodically corrugated waveguide with circular cross-section, with inner radius a and outer radius b. We model the wall roughness as a series

---

CP581, *Physics of, and Science with, the X-Ray Free-Electron Laser,* edited by S. Chattopadhyay et al.

of periodic (with period L) obstacles of height h ($h = b - a$) and thickness t (see figure 1 and 2). The charge travels along the z-axis; we assume $t \ll L$, $L \ll \lambda$ and the ohmic losses in the material negligible.

The periodicity of the geometry along the z-axis allows the use of Floquet's theorem which implies a field solution independent of the period L (obtained from a

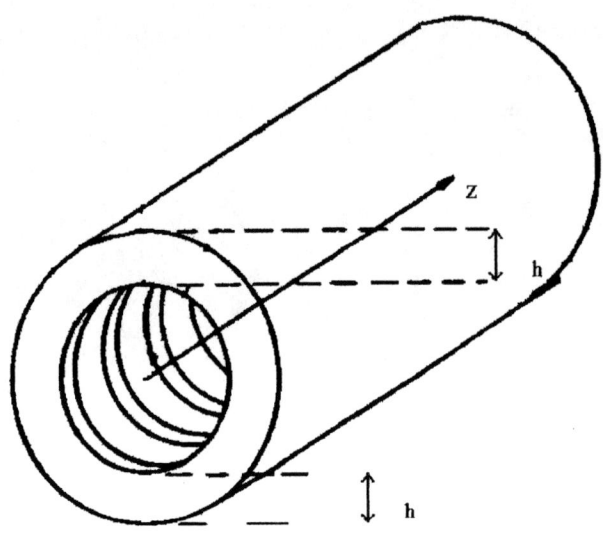

FIGURE 1. Relavant geometry.

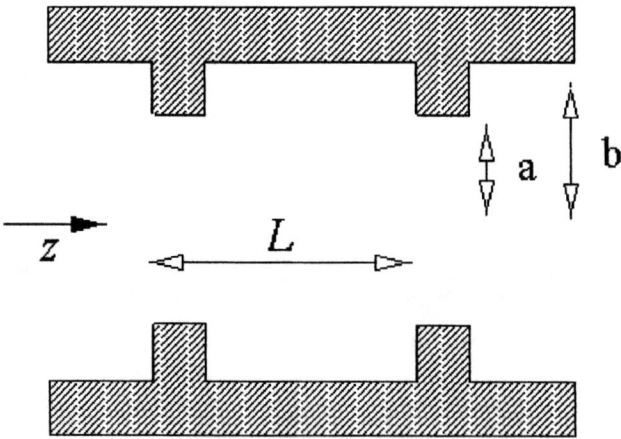

FIGURE 2. Schematic view of the waveguide and notation adopted.

34

single cell). The steps are the following: at first we solve the homogeneous problem, finding the modes propagating in the waveguide and their features (the dispersion equation, the cut-off frequency and the frequency where the synchronous wave is excited). The dispersion relation is found by applying the continuity conditions for the field components over the boundary between the slot (the space inside the corrugation) and the internal region of the waveguide. The field inside the waveguide is considered as generated by the magnetic and the electric Hertz potentials along the z-axis. Then we apply the Lorentz reciprocity principle, including the charge as an impulsive source, finding the coefficients used to express the electric field along the z-axis.

RESULTS

A   The Homogeneous Problem

The electromagnetic fields inside the corrugation are considered to be those due to propagating radial modes; higher-order evanescent modes are considered negligible, this assumption is justified under the hypothesis that the wavelength is much greater than the distance between two corrugations ($\lambda \gg L$). The components of the electromagnetic inside the corrugation field are:

$$E_r^C = 0 \tag{1}$$

$$E_\phi^C = 0 \tag{2}$$

$$E_z^C = \sum_n [C_n J_n(k_0 r) + D_n Y_n(k_0 r)] \cos(n\phi) \tag{3}$$

$$H_r^C = \sum_n \frac{n}{j\omega\mu r} [C_n J_n(k_0 r) + D_n Y_n(k_0 r)] \sin(n\phi) \tag{4}$$

$$H_\phi^C = \frac{k_0}{j\omega\mu} \sum_n [C_n J_n'(k_0 r) + D_n Y_n'(k_0 r)] \cos(n\phi) \tag{5}$$

$$H_z^C = 0 \tag{6}$$

where $e^{j(\omega t - \beta_n' z)}$ is assumed and

$$\beta_n' = \sqrt{k_0^2 - k_t^2} \tag{7}$$

35

β'_n is the hybrid-mode propagation constant, k_0 is the free-space propagation constant and k_t is the transverse propagation constant. The fields inside the waveguide is considered as generated by the Hertz potentials along the z-axis:

$$\Pi_{ez} = \sum_n A_n J_n(k_t r) \cos(n\phi) e^{j(\omega t - \beta'_n z)} \tag{8}$$

$$\Pi_{mz} = \sum_n B_n J_n(k_t r) \sin(n\phi) e^{j(\omega t - \beta'_n z)} \tag{9}$$

which are related to the field by the relation:

$$\mathbf{E} = -j\omega\mu\nabla \times \mathbf{\Pi_m} + (\mathbf{k^2} + \nabla\nabla\cdot)\mathbf{\Pi_e} \tag{10}$$

$$\mathbf{H} = j\omega\epsilon\nabla \times \mathbf{\Pi_e} + (\mathbf{k^2} + \nabla\nabla\cdot)\mathbf{\Pi_m} \tag{11}$$

where

$$\mathbf{\Pi_m} = \Pi_{mz}\mathbf{z_0} \qquad \mathbf{\Pi_e} = \Pi_{ez}\mathbf{z_0} \tag{12}$$

The potential Π_{ez} generates TM_z modes and the potential Π_{mz} TE_z modes. The superposition of both kind of modes gives origin to hybrid modes. So the expressions of the components of the hybrid modes in the internal region of the waveguide are given by equations 10 and 11 using the relations 8 and 9:

$$e_{rn} = \left[-jB_n\omega\mu\frac{n}{r}J_n(k_t r) - jA_n\beta'_n k_t J'_n(k_t r) \right] \cos(n\phi) e^{j(\omega t - \beta'_n z)} \tag{13}$$

$$e_{\phi n} = \left[jB_n\omega\mu k_t J'_n(k_t r) + jA_n\beta'_n\frac{n}{r}J_n(k_t r) \right] e^{j(\omega t - \beta'_n z)} \tag{14}$$

$$e_{zn} = A_n k_t^2 J_n(k_t r) \cos(n\phi) e^{j(\omega t - \beta'_n z)} \tag{15}$$

$$h_{rn} = \left[-jB_n\beta'_n k_t J'_n(k_t r) - jA_n\omega\epsilon\frac{n}{r}J_n(k_t r) \right] e^{j(\omega t - \beta'_n z)} \tag{16}$$

$$h_{\phi n} = \left[-jB_n\beta'_n\frac{n}{r}J_n(k_t r) - jA_n\omega\epsilon k_t J'_n(k_t r) \right] \cos(n\phi) e^{j(\omega t - \beta'_n z)} \tag{17}$$

$$h_{zn} = B_n k_t^2 J_n(k_t r) \sin(n\phi) e^{j(\omega t - \beta'_n z)} \tag{18}$$

Applying the boundary condition at $r = b$ (the electric field in the perfectly conducting wall vanishes) from equation 3 we find:

36

$$E_z^C(b) = 0 \implies D_n = -\frac{C_n J_n(k_0 b)}{Y_n(k_0 b)} \tag{19}$$

and imposing the continuity of the tangential components of the field at $r = a$:

$$E_z^C = E_z \qquad H_\phi^C = H_\phi \qquad for \qquad r = a \tag{20}$$

that is, using equation 19

$$C_n J_n(k_0 a) - \frac{J_n(k_0 b)}{Y_n(k_0 b)} C_n Y_n(k_0 a) = A_n k_t^2 J_n(k_t a) \tag{21}$$

$$\frac{k_0}{j\omega\mu} \left[C_n J_n'(k_0 a) - \frac{J_n(k_0 b)}{Y_n(k_0 b)} C_n Y_n'(k_0 a) \right] = -j B_n \beta_n' \frac{n}{a} J_n(k_t a) - j A_n \omega \epsilon k_t J_n'(k_t a) \tag{22}$$

from equations 21 and 22 we find the dispersion relation for the hybrid modes:

$$\frac{J_n'(k_0 a) Y_n(k_0 b) - J_n(k_0 b) Y_n'(k_0 a)}{J_n(k_0 a) Y_n(k_0 b) - J_n(k_0 b) Y_n(k_0 a)} = \frac{k_0}{k_t} \frac{J_n'(k_t a)}{J_n(k_t a)} - j \frac{\beta_n'^2 n^2}{a^2 k_t^3 k_0} \frac{J_n(k_t a)}{J_n'(k_t a)} \tag{23}$$

The modes of interest are the TM_{0m}: the TE_z and the TM_{nm} (with $n \neq 0$) modes do not give any contribution in the application of the reciprocity principle.
The dispersion relation 23 for $n = 0$ becomes:

$$\frac{J_0'(k_0 a) Y_0(k_0 b) - J_0(k_0 a) Y_0(k_0 b)}{J_0(k_0 a) Y_0(k_0 b) - J_0(k_0 b) Y_0(k_0 a)} = \frac{k_0}{k_t} \frac{J_0'(k_t a)}{J_0(k_t a)} \tag{24}$$

In the hypothesis of small corrugations $(a \to b)$ the cut-off frequency is found to be:

$$f_{co} = \frac{c}{2\pi} \left(\frac{\xi_{01}}{a+h} \right) \tag{25}$$

where ξ_{01} is the first zero of the Bessel function of first kind and order zero. It has to be noted that, for $h \to 0$, 25 tends to the cut-off frequency of the TM_{01} mode in the smooth waveguide of radius a, as expected. In the same hypothesis of small corrugations we found the frequency where the synchronous wave is excited (crossing frequency):

$$\bar{f}_{cr} = \frac{c}{2\pi} \frac{\xi_{01}}{\sqrt{ah}} \tag{26}$$

It shows the typical behaveour of proportionality to $1/\sqrt{h}$. In figures 3 and 4 are reported the Brillouin diagrams when $h/a = 0.1$ and when $h/a = 0.8$.
The red line is the dispersion curve for the TM_{01} mode in the corrugated waveguide and the blue one is the straight line $\beta_0' = k_0$. For $\beta_0' > k_0$ the wave is slow and can be synchronous with the charge. It has to be noted that for the biggest value of h (figure 4), the slow wave will be excited for a smaller frequencies.

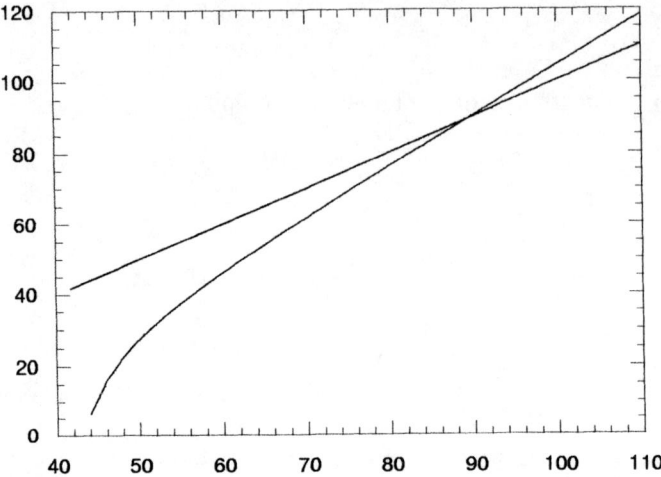

FIGURE 3. Brillouin diagram for a circular cross-section waveguide of radius a with periodic corrugations of depth h and $h/a = 0.1$.

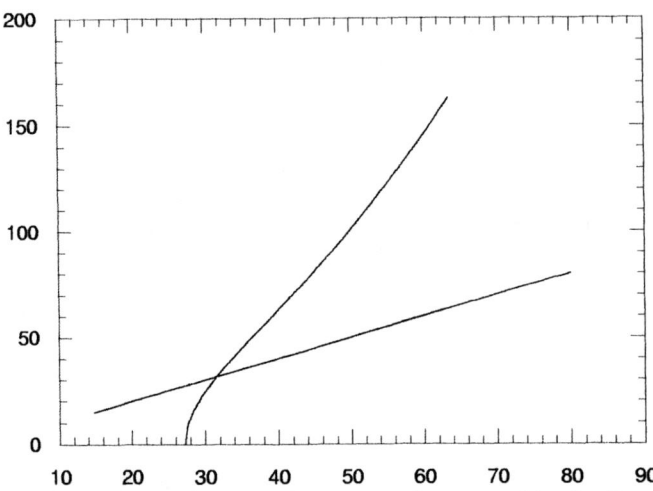

FIGURE 4. Brillouin diagram for a circular cross-section waveguide of radius a with periodic corrugations of depth h and $h/a = 0.8$.

B   Including the Sources

Once derived the modes of structures, having solved the homogeneous problem, the field genereted by a point charge can be found by means of the Lorentz reciprocity principle [3]:

$$\int_{S} \left(\mathbf{E}_n^{\pm} \times \mathbf{H} - \mathbf{E} \times \mathbf{H}_n^{\pm} \right) \cdot \mathbf{n} dS = \int_{V} \mathbf{J} \cdot \mathbf{E}_n^{\pm} dV \qquad (27)$$

38

The sign $+$ is for the wave traveling along $+z$ and the sign $-$ is for the wave traveling along $-z$. The current density of a point charge traveling on-axis, $\mathbf{J}$, used in the reciprocity principle, is modeled as an impulsive source

$$\mathbf{J}(r, \phi, z; \omega) = q \frac{\delta(r)}{r} \delta(\phi) e^{-j\frac{z}{v_p}\omega} \mathbf{z_0} \tag{28}$$

where $\mathbf{z_0}$ is the unit vector along z-axis and q is the charge. The electromagnetic field can be express using the modal expansion [3]:

$$\mathbf{E} = \sum_n (a_n \mathbf{E^+}_n + b_n \mathbf{E^-}_n) \tag{29}$$

$$\mathbf{H} = \sum_n (a_n \mathbf{H^+}_n + b_n \mathbf{H^-}_n) \tag{30}$$

where

$$\mathbf{E^+}_n = (\mathbf{e_{tn}} + e_{zn}\mathbf{z_0})e^{-j\beta_{zn}z} \qquad \mathbf{E^-}_n = (\mathbf{e_{tn}} - e_{zn}\mathbf{z_0})e^{+j\beta_{zn}z} \tag{31}$$

$$\mathbf{H^+}_n = (\mathbf{h_{tn}} + h_{zn}\mathbf{z_0})e^{-j\beta_{zn}z} \qquad \mathbf{H^-}_n = (-\mathbf{h_{tn}} + e_{zn}\mathbf{z_0})e^{+j\beta_{zn}z} \tag{32}$$

The coefficients a_n and b_n are found applying the Lorentz principle 27. For $n = 0$, the present case, they are given by:

$$a'_0 = -\frac{q\delta(\frac{k_0}{\beta} - \beta'_0)}{2\pi\omega\epsilon\beta'_0 a^2 F(k_t a)} \tag{33}$$

$$b'_0 = \frac{q\delta(\frac{k_0}{\beta} + \beta'_0)}{2\pi\omega\epsilon\beta'_0 a^2 F(k_t a)} \tag{34}$$

where

$$F(k_t a) = J_1^2(k_t a) - J_0(k_t a)J_2(k_t a) \tag{35}$$

and $J_0(k_t a)$, $J_1(k_t a)$ and $J_2(k_t a)$ are the Bessel functions of first kind and order 0, 1 and 2, respectively.

So it becomes possible to find the expression of the electric field in the frequency domain along the z-axis using equation 29 for $n = 0$

$$E_z = a_0 e_{z0}^+ - b_0 e_{z0}^- \tag{36}$$

using the coefficients expressed by 33 and 34 36 becomes

$$E_z(r, \phi, z; \omega) = -\frac{q}{\bar{\omega}_{cr}\epsilon\beta'_0 a^2 F(k_t a)} k_t J_0(k_t r) \left[\delta(\frac{k_0}{\beta} - \beta'_0) + \delta(\frac{k_0}{\beta} + \beta'_0) \right] e^{-j\beta'_0 z} \tag{37}$$

where $\bar{\omega}_{cr}$ is the crossing frequency (expressed by 2π times the equation 26), ϵ is the dielectric constant in free-space. In the hypothesis of small corrugation ($h \to 0$) and for ultrarelativistic particles ($\beta \to 1$) the electric field in the time domain is

$$E_z(z;\tau) = -\frac{8qZ_0ch}{(\xi_{01})^2\pi a^3}\cos(\bar{\omega}_{cr}\tau)e^{-j\beta_0'z} \tag{38}$$

where c is the light velocity in free-space and Z_0 is the free-space characteristic impedence. From equation 38 it can be noted that E_z has a π phase difference with the charge (the - sign), meaning that it is a decelerating field and that the height of the corrugation h fixes not only the crossing frequency, but also the field amplitude through the factor h/a^3.

C Longitudinal Coupling Impedance and Wake Function

Following the standard definition of the longitudinal coupling impedance per unit length [4]:

$$\frac{\partial Z(\omega)}{\partial z} = -\frac{1}{q}E_z\left(x = 0, y = 0, z, \omega\right)e^{j\omega z/c}, \tag{39}$$

from equation 37 we get

$$\frac{\partial Z_z(\omega)}{\partial z} = \frac{4Z_0ch}{(\xi_{01})^2\pi a^3}\left[\delta(\omega - \bar{\omega}_{cr}) + \delta(\omega + \bar{\omega}_{cr})\right] \tag{40}$$

Again from the definition [4], it is easy to get the longitudinal wake function per unit length

$$\frac{\partial w(\tau)}{\partial z} = -\frac{E_z(z;\tau)}{q}e^{j\omega z/c}. \tag{41}$$

From equation 38:

$$\frac{\partial w_z(z;\tau)}{\partial z} = \frac{8Z_0c}{(\xi_{01})^2\pi}\frac{h}{a^3}\cos(\bar{\omega}_{cr}\tau) \tag{42}$$

Both the longitudinal coupling impedence 40 and the longitudinal wake function 42 are proportional to the depth height h.

D The LCLS Case

As an example we report the application of the theory developed above to the case of the LCLS undulator [5]. We consider the case of a rectangular bunch, of temporal dimension $2T$, where T is given by $T = \frac{\sqrt{3}}{c}\sigma_l$, being σ_l $(= 15 \ \mu m)$ the longitudinal dimension of the bunch.

We find that the crossing frequency is given by

$$\bar{f}_{cr} = 2.29 \cdot 10^{11} \frac{1}{\sqrt{h}} \qquad [Hz] \tag{43}$$

and the amplitude of the wake function per unit length is given by

$$w_0' = 3.2017 \cdot 10^{18} h \qquad [V/Cm] \tag{44}$$

Let us focus our attention on the energy spread, that is given by [6]

$$\frac{\Delta E^{rms}}{E_0} = \frac{w_0' DQ}{2E_0} \left[\frac{1}{2(\bar{\omega}_{cr}T)^2} \left(1 - \frac{\sin(4\bar{\omega}_{cr}T)}{4\bar{\omega}_{cr}T}\right) - \left(\frac{\sin(\bar{\omega}_{cr}T)}{\bar{\omega}_{cr}T}\right)^4 \right]^{\frac{1}{2}} \tag{45}$$

where $E_0 = 14.35 \ GeV$ is the total energy of the electron beam, $D = 112 \ m$ is the total length of the path followed by the beam and $Q = 0.1 \ nC$ is the bunch charge. In figure 5 is reported the energy spread vs. $h \ [m]$.

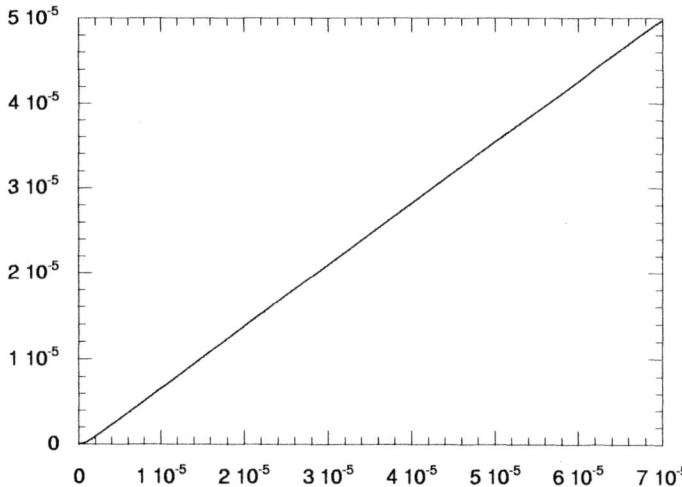

FIGURE 5. Energy spread for the circular cross-section waveguide vs. h.

The design parameters require the energy spread to be less than $5 \cdot 10^{-4}$, which is verified for value of h of the order of tens of microns.

It is easy to calculate the loss factor that is given by [4]

$$K = \frac{w_0'}{2} \left[\frac{\sin(\bar{\omega}_{cr}T)}{\bar{\omega}_{cr}T}\right]^2. \tag{46}$$

In figure 6 the loss factor is reported vs. $h \ [m]$.

41

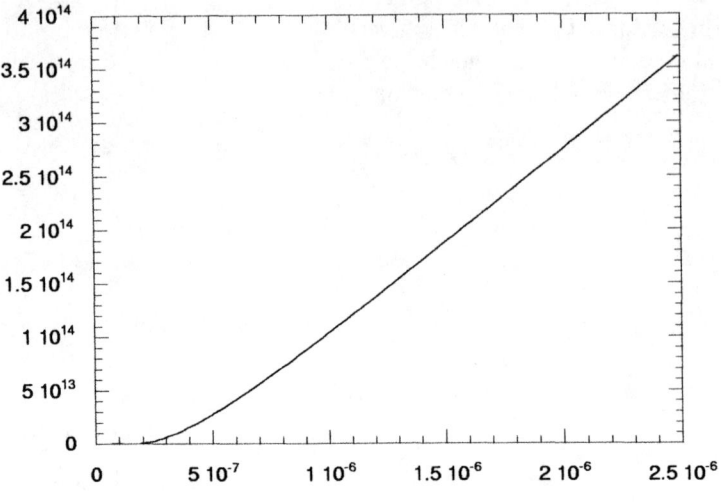

FIGURE 6. Loss factor [V/C] vs. h [m].

CONCLUSIONS

We have derived the longitudinal wake due to a periodic corrugation in a circular beam pipe. The amplitude of sinusoidal the wake function is proportional to h and the crossing frequency (where the slow wave synchronous with the beam can be excited) is proportiional to $1/\sqrt{h}$. For $h \to 0$ the crossing frequency goes to infinity and the amplitude of the wake function vanishes. From our results it is possible to say that, in the case of the LCLS undulator, a beam pipe roughness of the order of magnitude of the tens of microns is accetable.

APPENDIX

Solution of the Dispersion Relation

In the hypothesis of small corrugations and from the definition of Wronskian, it is possible to use the following approximations [7]

$$J_0'(k_0 a)Y_0(k_0 b) - J_0(k_0 a)Y_0'(k_0 b) = -\frac{2}{\pi k_0 a} \tag{47}$$

$$J_0(k_0 a)Y_0(k_0 b) - J_0(k_0 b)Y_0(k_0 a) = \frac{2h}{\pi a} \tag{48}$$

so the dispersion relation becomes

$$\frac{k_t}{k_0} \frac{J_0(k_t a)}{J_0'(k_t a)} = -k_0 h \tag{49}$$

Considering that $k_0^2 = k_t^2 + k_z^2$, and that at cut-off $k_z = 0$, $k_0 = k_t$, we get:

$$\frac{J_0(k_0 a)}{J_0'(k_0 a)} = -k_0 h \tag{50}$$

solving this equation for k_0

$$k_0 = \frac{k_0^S}{a + h} a \tag{51}$$

where k_0^S is the propagation constant for the smooth waveguide. It is immedidiate to find the cut-off frequency from the propagation constant expressed by 51.

For the smooth waveguide the dispersion reletion is given by:

$$J_0(k_t a) = 0 \quad \Longrightarrow \quad k_t a = \xi_{om} \tag{52}$$

where ξ_{0m} is the m-th zero of the Bessel function del primo of the first kind of oreder 0.

Being x the distance from the zero ξ_{0m} and making $x \to 0$ it is possible to write:

$$J_0'(k_t a) = J_0'(\xi_{0m} - x) \to -J_1(\xi_{0m}) - \frac{x}{\xi_{0m}} J_1(\xi_{0m}) \tag{53}$$

$$J_0(k_t a) = J_0(\xi_{0m} - x) \to J_1(\xi_{0m}) x \tag{54}$$

Form equations 53, 54, 47 and 48 it is possible to get an other approximated expression of the dispersion relation

$$-\frac{1}{x} \left(1 + \frac{x}{\xi_{0m}}\right) \frac{k_0}{k_t} = -\frac{1}{k_0 h} \tag{55}$$

Solving this second order equation in x

$$x = \frac{k_0^2 h a}{\xi_{0m}} \tag{56}$$

we get

$$k_t a = \xi_{0m} - \frac{(k_0 a)^2}{\xi_{0m}} \frac{h}{a} \tag{57}$$

Imposing the crossing condition $k_t = 0$ we get the propagation constant at the crossing frequency

$$\bar{k}_0 = \frac{\xi_{0m}}{\sqrt{ah}} = \frac{2.40}{\sqrt{ah}} = \frac{\bar{\omega}}{c} \tag{58}$$

it is immediate to get the crossing frequency from this equation.

Calculation of the Coefficients for the Modal Expansion

The Lorentz reprocity principle assumes the expression

$$\sum_m b_m A_{nm} - \sum_m a_m B_{nm} = \int_V \mathbf{J} \cdot \mathbf{E}_n^+ dV \tag{59}$$

$$\sum_m b_m C_{nm} - \sum_m a_m D_{nm} = \int_V \mathbf{J} \cdot \mathbf{E}_n^- dV \tag{60}$$

where the coefficients A_{nm}, B_{nm}, C_{nm} and D_{nm} are related to the transerse of the the n-th mode. In the case under examination $n = m$, equation 60 becomes

$$a_n A_{nn} = \int_V \mathbf{J} \cdot \mathbf{E}_n^- dV \tag{61}$$

and

$$A_{nn} = 2 \int_S (\mathbf{e}_{tn} \times \mathbf{h}_{tn}) \cdot z_0 dS \tag{62}$$

that for $n = 0$ is

$$A_{00} = -2\omega\epsilon\beta_0' k_t^2 \int_0^a [J_0'(k_t r)] \, r \, dr \int_0^{2\pi} d\phi \tag{63}$$

solving the integral we found

$$A_{00} = 2\pi\omega\epsilon\beta_0' k_t^2 a^2 F(k_t a) \tag{64}$$

From equations 61 and 64 it is easy to find the expression 33. Anologously we found the espression for the coeficient 34 from 59.

REFERENCES

1. G. H. Bryant, Proc. IEE **116**, 203 (1969).
2. L. Palumbo et al., Wake Fields Effects due to Surface Roughness, *VIII International Workshop on Linear Colliders, Frascati*, 1999.
3. R. E. Collin, *Field Theory of Guided Waves*, 2nd ed. (Oxford Univ. Press, Oxford, 1995).
4. L. Palumbo V.G. Vaccaro, M. Zobov, in *Proc. CERN Accelerator School: Advanced Accelerator Physics Course, Rhodes, 1993*, No. 95-06 in *CERN Yellow Report*, edited by S. Turner (European Lab. for Particle Physics, Geneva, Switzerland, 1995), pp. 331–390.
5. LCLS Design Study Group, SLAC Report No. SLAC-R-521, 1998.
6. K.L.F. Bane and A. Novokhatskii, SLAC Report No. SLAC-AP-117, 1999
7. C. A. Balanis, *Advanced Engineering Electromagnetics*, (Wiley Sons, New York, 1989).

Plasma-based studies on 4th Generation Light Sources

R. W. Lee[a], H. A. Baldis[b], R. C. Cauble[a], and O. L. Landen[a], J. S. Wark[c], A. Ng[d], S. J. Rose[e], C Lewis[f], D. Riley[f], J.-C. Gauthier[g], and P. Audebert[g]

[a]L-399 Lawrence Livermore National Laboratory, Livermore, CA
[b]Institute for Laser Science and Applications / University of California, Davis
[c]Department of Physics, Clarendon Laboratory, Oxford University, Oxford, UK
[d]Department of Physics & Astronomy, University of British Columbia, Canada
[e]Central Laser Facility, Rutherford Appleton Laboratory, Chilton, UK
[f]Physics Department, Queen's University Belfast, Belfast, Northern Ireland
[g]LULI UMR 7605 Ecole Polytechnique-CNRS/CEA-Université Paris VI France

Abstract: The construction of a short pulse tunable x-ray laser source will be a watershed for plasma-based and warm dense matter research. The areas we will discuss below can be separated broadly into warm dense matter (WDM) research, laser probing of near solid density plasmas, and laser-plasma spectroscopy of ions in plasmas.

The area of WDM refers to that part of the density-temperature phase space where the standard theories of condensed matter physics and/or plasma statistical physics are invalid. Warm dense matter, therefore, defines a region between solids and plasmas, a regime that is found in planetary interiors, cool dense stars, and in every plasma device where one starts from a solid, *e.g.*, laser-solid matter produced plasma as well as all inertial fusion schemes.

The study of dense plasmas has been severely hampered by the fact that laser-based methods have been unavailable. The single most useful diagnostic of local plasma conditions, *e.g.*, the temperature (T_e), the density (n_e), and the ionization (Z), has been Thomson scattering. However, due to the fact that visible light will not propagate at electron densities, $n_{e,} \geq 10^{22}$ cm^{-3} implies dense plasmas can not be probed. The 4th generation sources, LCLS and Tesla will remove these restrictions.

Laser-based plasma spectroscopic techniques have been used with great success to determine the line shapes of atomic transitions in plasmas, study the population kinetics of atomic systems embedded in plasmas, and look at redistribution of radiation. However. the possibilities end for plasmas with $n_e \geq 10^{22}$ since light propagation through the medium is severely altered by the plasma. The entire field of high Z plasma kinetics from laser produced plasma will then be available to study with the tunable source.

I. BACKGROUND

Since the late 1960's plasma-based research has moved toward higher density regimes. The advent of laser-produced plasmas and laser-based plasma diagnostics have fueled interest in the formation of plasmas at densities nearing solid density.

CP581, *Physics of, and Science with, the X-Ray Free-Electron Laser,* edited by S. Chattopadhyay et al.
© 2001 American Institute of Physics 0-7354-0022-9/01/$18.00

There are two separate areas where the 4th generation sources, LCLS and Tesla, can play a critical role in moving this field substantially forward. The first is in the area of warm dense matter research, while the second is in the area of plasma spectroscopic techniques.

We note that whether we are interested in creating warm dense matter, performing Thomson scattering, or probing a plasma the LCLS/Tesla capability provides a major advance on any capability that exists with 3rd generation sources. The key to the advance is the tunable, narrow band x-ray source with very short pulse duration. Since the individual bunch photon intensity is the essential quantity for all the plasma-based research, the comparison of the LCLS/Tesla to current synchrotron sources is best summarized by comparing peak spectral brightness. Indeed, one finds a 10 order of magnitude enhancement that will make the LCLS/Tesla a most promising source for plasma based research. The utility of the high repetition rate of other sources, *e.g.,* APS or ESRF, are not useful here since we require a single photon pulse to either heat, scatter, or probe matter that is transient. Indeed, to create solid matter that is at a temperature greater than 1 eV temperature while not expanding requires the LCLS/Tesla.

Further, to measure the Thomson signal in a dense plasma or the warm dense matter regime requires a Thomson probe, here the 4th generation, with temporal duration that is short compared to the evolution of the system but long enough to probe the plasma collective modes. The LCLS with the nominal 200 fs pulse duration will be able to probe the electron feature, arising from the collective behavior of the electrons, for an electron density of 10^{22} cm^{-3} one finds that 1 fs is sufficient for collective response. However, the ion collective behavior will take on the order 100 fs leading to possibility that the collective ion acoustic modes will not be sampled appropriately. The evolution of the system will proceed on the order of a picosecond. This indicates that any probe of the system must have a pulse duration on the ps time scale.

Finally, the spectroscopic probing of high energy density plasmas requires a short pulse high energy source. For the source to be useful as a spectroscopic probe requires a spectrally tunable source for which the number of photons per mode must be on the order of unity. Photons per mode measure of the probe's ability to dominate a radiative transition that is required to provide observable signal and due to the high peak brightness the 4th generation sources will provide this capability.

A. Warm Dense Matter

With a short duration pulse containing a substantial number of high energy photons one can generate solid matter at temperatures of ≤ 10 eV, *i.e.,* warm dense matter. The interest in the warm dense matter regime arises because in dense plasmas the atoms and/or ions will start to behave in a manner that is intrinsically coupled to the plasma. That is, the plasma starts to exhibit long- and short-range order due to the correlating effects of the atoms/ions. This intriguing regime where the plasma can no longer be considered a thermal bath and the atoms are no longer well described by their isolated atom behavior provides a tremendous challenge to researchers. In the limit of dense cool plasmas one obviously arrives at the threshold of condensed matter. Here the problem has changed from a perturbative approach to ground-state methods where complete renormalization of the atom/ion and it environment is essential.

From the prospective of plasma studies the defining quantity is the coupling parameter Γ, i.e., the ratio of the inter-atomic potential energy to the thermal energy given by the equation:

$$\Gamma = \frac{Z^2 e^2}{r_0 kT} \text{ with } r_0 = \left(\frac{3Z}{4\pi n_e} \right)^{1/3}$$

where Z is the ion charge and r_o is the interparticle spacing given in terms of the electron density n_e.

The regions of interest span the density-temperature phase space going from modestly coupled ($\Gamma \leq 1$) to strongly coupled ($\Gamma > 1$), while bridging the transition regimes between solid to liquid to plasma.

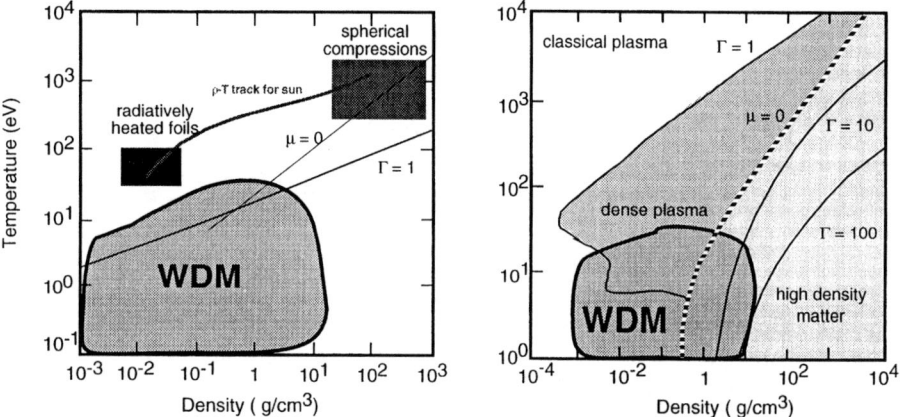

Figure 1. The temperature-density phase diagram for hydrogen on the left and aluminum on the right. The relevant regimes are noted, as are the various values of the coupling Γ. The regions greatest uncertainty are roughly noted by the black outlined areas. Also indicated is the region where degeneracy will become important: it is the region to the right of the line where the chemical potential $\mu = 0$.

In figure 1 above we show the region of the temperature-density plane where warm dense matter studies are important. Here we show the temperature (T) in eV versus the density (ρ) in g/cm^3 both for hydrogen, a low Z element, and aluminum, a moderate Z element. The region where the theoretical uncertainties are largest are those where the standard theoretical approaches fail and experiments are exceedingly difficult. The difficulty arises theoretically from the fact that this is a regime where there are no obvious expansion parameters, as the usual perturbation expansions in small parameters used in plasma phase theories are no longer valid. Further, there becomes an increased importance on density-dependent effects, e.g., pressure ionization, as the surroundings starts to impinge on the internal structure of the ion or atom. Experimentally the study of warm dense matter is difficult, as the isolation of samples in this regime is complicated. Indeed, although the plasma evolution of *every ρ-T* path that starts from the solid phase goes through this regime and plays an important role in its evolution, trying to isolate warm dense matter remains a major challenge.

It has been exceedingly difficult to perform experiments in the warm dense matter regime, which is, simply, why we know so little about it. As a first step, one must create a well-characterized warm dense matter state; the second is to gain information on the state through experiments. The first step has been the problem: warm dense matter is not a limiting case of matter, *e.g.*, high- or low-temperature. When created in a laboratory environment, it does not tend to remain in a specified thermodynamic state for very long, making characterization difficult. The only other imaginable method to produce the kind of warm dense matter of interest here might be to use sub-30-fs laser pulses on sub-100-Å-thick foils and perform thermodynamic measurements on a few-fs timescale over extremely small spatial dimensions. To be able to do this on comparatively macroscopic samples with 4th generation sources will be a boon.

B. Plasma Spectroscopic Studies

There is great interest in the higher temperature dense plasma regime. Here the problem arises from the production of high temperature plasmas at electron densities in excess of 10^{22} cm^{-3}. In any experiment where a high intensity, *e.g.*, $I \geq 10^{12}$ W/cm^2, laser irradiates a solid target there will be a region of the solid that is hot and near solid densities. Lasers with wavelengths > 0.25 μm do not directly heat the solid as they can not propagate beyond the critical electron density $\sim 10^{21}$ cm^{-3} x $(1 \mu m / \lambda_{laser})^2$ [8]; however, heat flow from the surface efficiently generates the hot dense medium.[8] The spectroscopic information derived from these plasmas provides, on the one hand, diagnostic information about the plasma itself, while on the other hand we can investigate, using spectroscopy, our understanding of the mechanism at play in the creation of the plasma and the interaction of the atoms/ions with the plasma in which it is embedded. Here the LCLS/Tesla will provide two related and intriguing possibilities. First, there is the possibility to perform Thomson scattering on plasmas at solid density.[9,10] Second, we can explore laser pump-probe techniques for high density plasmas that have been used in low densities plasmas to measure line shapes, observe radiation redistribution, and determine the kinetics processes.[11,12]

1. Thomson Scattering

Thomson scattering provides an *in situ* measurement of the temperature, density, charge state, and collective behavior of the plasma. Indeed, the Thomson scattering diagnostic is directly related to the dynamic structure factor, $S(k,\omega)$, of the plasma and thus provides insight into the theoretical predictions from different theories. It is fair to say that in recent years each effort at diagnosing a higher density plasma, *i.e.*, higher than 10^{20} cm^{-3}, using Thomson scattering has led to new and important discoveries.[13] These experiments have, of course, been few since the constraints on the experiments are substantial. Here we believe that the 4th generation sources will provide a major advance in diagnosing dense plasmas. This is clearly a complement to the concept of creating warm dense matter, as Thomson scattering can provide a diagnostic of the warm dense matter conditions. However, the preconditions for the interpretation of the scattering data is that there is a valid theoretical model for the $S(k, \omega)$ in the high density regime, and this in itself will be a challenge. The tunable nature of the x-ray source, the high energy, bandwidth, the short pulse duration and, importantly, the very high peak photon flux make this source the only one that can address the Thomson scattering of transient plasmas.

2. Laser Pump Probe Techniques

The mechanisms involved in the formation of a plasma and the details of the kinetics processes can be illuminated by using a laser as a pump to selectively populate levels and thus redistribute radiation. In a particularly intriguing possibility one will be able to study the formation of laboratory x-ray lasers that currently depend on kinetics processes.[14] Thus, one could disentangle the plasma production from the inversion-forming processes that lead to the x-ray lasing. It is clear that numerous aspects of plasma spectroscopy have been severely constrained by a lack of data. The 4th generation sources will provide a substantial improvement in the development of our understanding of intrinsic line shape formation, level shifts, radiation transfer, and detailed kinetics processes.

In both of these areas the LCLS/Tesla will provide information that would not be obtainable with any other source. The combination of the short pulse length, the tunable wavelength, the repetition rate, and the energy per pulse will make the data derived from these plasma-based experiments a major advance in our knowledge in this area.

II. EXPERIMENTS
A. Creating and Probing Warm Dense Matter

The first scientific project comes in the area of warm dense matter research. This regime is accessed in all laboratory experiments where one creates a plasma from solid or near solid density targets; however, it is difficult to study this part of the plasma creation process in isolation. Rapid temporal variations, steep spatial gradients, and uncertain energy sources lead to indecipherable complexity. Indeed, although there has been much interest in this regime, witnessed by the literature on strongly coupled plasmas, there has been little progress.[15] The interest generated in laboratory experiments is mirrored in the astrophysical literature where the warm dense matter regime is found, for example, in the structural formation of large planets and brown dwarfs.[16,17,18,19,20]

The fact that the LCLS will allow the creation and probing of the warm dense matter regime in the laboratory, as discussed briefly below, will provide a set of data that will spark the field. The idea is simple but the impact will be vast, as the data obtained in the generation of the warm dense matter along an isochore, *i.e.*, a track of constant density, with subsequent probing along the release isentrope, *i.e.*, a track of constant entropy, will be unique and critically important for progress in the field. The importance of this data derives from the fact that to date the only possible method of generating warm dense matter is by shocking the material. The shock method provides information along the principal Hugoniot, that is, the locus of points in the pressure-density space that are accessed by a single shock – one point for each shock. Although this has been quite useful, it is a very limited set of data providing little information on the general behavior in the warm dense matter regime. Indeed, the amount of data that is currently available is so proscribed that one finds insufficient constraints on theoretical development. This can be illustrated by the curves in figure 2 where several predictions for an isochore of aluminum is presented in the temperature and density phase space. Note that the four theories shown in the figure *all* predict theoretical Hugoniots that fit the experimentally determined Hugoniots, but all differ rather dramatically along the isochore. As aluminum is the most studied material, figure 2 can be interpreted as the minimum degree of uncertainty in this field of research and makes obvious the need for experimental data in this regime.[21]

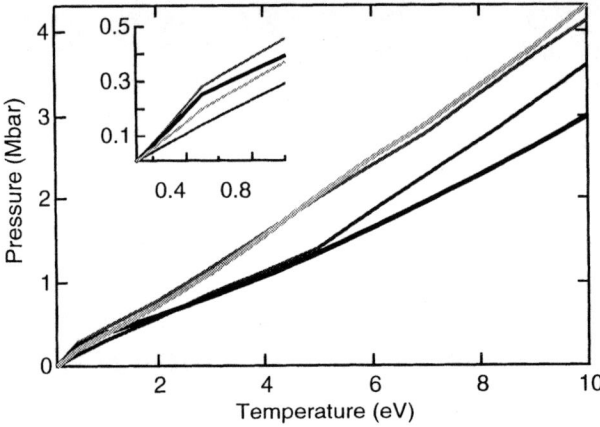

Figure 2. The isochore for aluminum in the warm dense matter regime for four theoretical models that all provide predictions that agree with the experimental point along a the principal Hugoniot. The inset shows the low pressure low density region expanded. Data derived from the LCLS will assist in motivating theoretical developments for this important regime.

B. Plasma Spectroscopic Studies

The 4th generation sources will be employed in plasma-based experiments to address the foundation of plasma creation in transition to hot dense matter providing a truly unique method to probe the spectroscopy of the hot dense matter. The probing of dense plasmas, whether these be warm or hot, will move to a new level of sophistication with the use of x-ray Thomson scattering. While, the active probing of the a hot dense plasma will be advanced by extending the methods of laser fluorescence spectroscopy that are employed in low density plasmas with visible lasers to high density using these x-ray laser sources.[22]

1. Thomson Scattering

One can extend the power of spectrally resolved Thomson scattering to the x-ray regime for direct measurements of the ionization state, density, temperature and the microscopic behavior of strongly coupled plasmas and warm dense solids. [10] This would be the first direct measurement of microscopic parameters of solid density matter, which could be used to properly interpret measurements of material properties such as thermal and electrical conductivity, equation of state (EOS) and opacity found in astrophysical environments as well as in virtually all plasma production devices.

Thomson scattering is characterized by the scattering parameter α, proportional to the ratio of the laser probe scale-length, λ_L, to the Debye length, λ_D, and the scattering angle Θ:

$$\alpha = \lambda_L/2\pi\lambda_D\sin(\Theta/2) = 1/k_L\lambda_D \sin(\Theta/2) \qquad (1)$$

For $\alpha < 1$, spectrally-resolved incoherent Thomson scattering provides information on the velocity v, hence temperature, and the directed flow of free electrons from the Doppler shifts experienced by scattered probe photons. For $\alpha > 1$, the collective scattering regime, the scattering is sensitive to temporal correlations between electron motion separated by more than a Debye length, hence the scattering is dominated by ion-acoustic and electron plasma wave resonances, the latter set by the Bohm-Gross dispersion relation.[8] The frequency shift of the resonance is dependent on density through the plasma frequency while the width of the resonances yields information on the wave damping rates. In the intermediate regime, $i.e.$, near $\alpha = 1$, the form of the high frequency electron plasma component depends strongly on both the electron temperature and density, providing a robust internal measurement of these basic plasma parameters, confirmed by spectroscopy.

In figure 3 we show the T-n_e regimes accessible by Thomson scattering with $\alpha = 2$ for various wavelength probes λ_L. Such Thomson scattering accesses regimes in which the Debye length is of order the probe wavelength. By switching from a visible laser probe at 5000 Å to an x-ray probe at, $e.g.$, 1.5 Å, we can effectively probe plasmas with Debye lengths of the order of the interparticle spacing or shorter (1 Å). Stated differently, for a given plasma temperature, we should be able to access a density that is 6-7 orders of magnitude higher than previously attempted. In particular, figure 3 shows that the solid density regime at $\sim 3 \times 10^{23}$ cm^{-3} crosses the strongly coupled plasma regime precisely where it is accessible by 1.5-12 Å Thomson scattering.

A schematic of the expected generic scattered spectrum features is shown in figure 4. Coherent scattering from tightly bound electrons (Z_{tb} per atom) will provide an unshifted peak at the probe wavelength whose intensity is proportional to Z_{tb}^2. Incoherent Compton scattering from weakly bound electrons (Z_{wb} per atom) should provide a second peak downshifted in energy by the order of $h\nu/mc^2$, with an intensity proportional to Z_{wb}. Thomson scattering from free electrons (Z_f per atom) should provide a dispersed spectrum centered on the Compton peak, with a spectrally integrated intensity varying as Z_f. The form of the spectrum will in general depend on

51

the free electron density, n_e, free electron and/or Fermi temperature T_{Fermi} and electron-ion collisionality v_{ei}. Hence, by spectrally resolving the scattered x-rays we would gain access, for the first time, to an unparalleled source of information on warm dense matter.

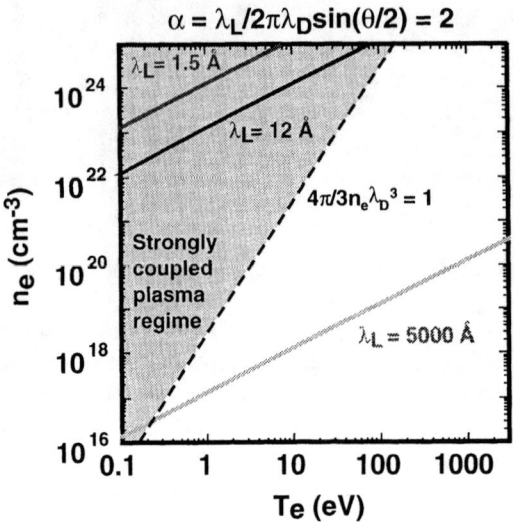

Figure 3. The temperature and electron density phase space indicating the non-ideal plasma regime by the shaded area. In the non-ideal region there is less than one particle per Debye sphere, thus leading to a breakdown in the Debye screening concept. The black dashed line indicates one particle per Debye sphere. Further, we indicate by solid lines the temperature and density at which the scattering parameter α is 2. One can easily observe that for the LCLS, whether at 1.5 Å or 12 Å, the dense regime is accessible, while for the nominal visible laser at 5000 Å one does not access the regime.

For example, we should be able to infer Z_f, Z_{tb}, and Z_{wb} from the relative importance of coherent, incoherent and free electron scattering contributions. This would allow us to discriminate between different ionization balance models used to define the EOS of these plasmas. [25] We should be able to infer the free electron temperature and density (and hence ionization state since the ion density is effectively hydrodynamically frozen) from the shape of the Thomson scattered spectrum for $\alpha \approx 1$, hence shedding further light on the equilibrium states of warm dense matter. We should be able to infer the plasma collisionality from the shape of the free electron spectral peak for $\alpha > 1$, hence allowing us to test the validity of various strongly-coupled statistical physics models.

We note that there is an ongoing effort to use laboratory produced soft x-ray lasers as Thomson scattering probes. Although there has been much progress recently reported it is important to review these efforts to understand fully the capabilities that will become available only when a 4th generation device is commissioned. There is now a laser-driven soft x-ray laser systems capable of producing megawatts of stimulated emission in the range of 100Å - 200Å from Ni-like collisional excitation schemes [26, 27].

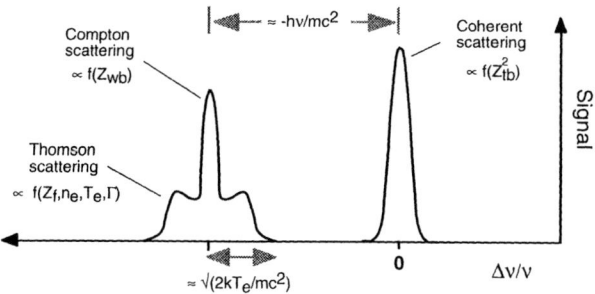

Figure 4. Schematic of spectrally-resolved x-ray scattering spectrum expected, with information provided by each feature noted. The definition of weakly bound and tightly bound electrons depend on their binding energy relative to the Compton energy shift. Those with binding energies (ionization potentials) less than the Compton shift are categorized weakly bound.

In a recent paper [28] calculations and theoretical analysis discussed the first application of a x-ray laser for probing hot, high-density plasmas ($n_e \geq 10^{23}$ cm^{-3}) using a Ni-like transient collisional excitation x-ray laser as a probe. Theoretical predictions were used to diagnose the electron temperature in short pulse (500 fs) laser produced plasmas. Necessary conditions were found for the probe intensity, spectral resolution of the spectrometer and sensitivity of the scattered x-ray detection system for the measurement of electron temperature.

In the analysis an x-ray probe is scattered from the high-density plasma produced by a subpicosecond laser pulse. The range of plasma parameters was $T_e \sim 50$ eV and $n_e \geq 10^{23}$ cm^{-3}, and the wavelength of the Thomson probe ($\lambda = 147$ Å) lead to values of $\alpha > 1$ for all scattering angles θ, thus remaining in the collective regime. The x-ray laser [26] produces a beam which is characterized by the coherence lengths on the order of 220 μm in the longitudinal direction and approximately 10 μm in the transverse direction. Both values are much longer than the correlation lengths of ion acoustic fluctuations produced in the plasma.

Using the expression for the dynamical form factor, $S(k, \omega)$, spatially averaged profiles of the Thomson scattering cross-sections have been calculated at different times. Early time scattering is characterized by the wide separation of ion-acoustic maxima due to larger temperature, but smaller scattering levels. At these early times, before the foil explosion, the effective scattering volume is smaller, reducing the overall scattered power. It was found that with an assumed instrumental resolution of $\Delta\lambda/\lambda = 10^{-4}$ the two ion-acoustic peaks may be differentiated and thus the T_e determined. Estimates of instrumental sensitivity show that approximately 70 photons are detected within an instrumental resolution channel when the solid angle collection and reflectivity of a large cylindrical optic, entrance slit transmission, multilayer grating reflectivity and typical back-thinned CCD quantum efficiency are included. Moreover, it was calculated that intensities near $I_p = 10^{12}$ W/cm^2 are required to probe these simulated plasmas, close to the current operating regime of the Ni-like transient x-ray laser system, *i.e.*, 2×10^{11} W/cm^2).

This effort is a step towards the future realization of a 4th generation Thomson scattering experiments. Although Thomson scattering using soft x-ray lasers can provide unique information from high-density plasmas, its application right now is severely limited to the microscopic plasmas created with high-power short pulse lasers. Further, the relatively large attenuation and refraction that occur in dense plasmas, as well as the inherently low shot rate will make these measurements difficult. Nevertheless, these experiments will prove invaluable to the application of x-ray Thomson scattering for the future sources.

We note in closing that there is the intriguing possibility of using part of the LCLS beam, for example the spontaneous background, to warm a sample and then use the LCLS to probe it with Thomson scattering. This is in distinction to the idea that we create the plasma using a standard long pulse laser system. The experiments using the LCLS as the heater and probe are much more demanding as the number of free electrons may decrease by at least an order of magnitude. Note that the number of free electrons is a matter of great controversy in this regime as the definition of "free" is open to debate. [29] Further, the need to delay the heater from the probe by x-ray optics creates addition losses of $\sim 10^{-4}$ yielding the total number of detected photons per shot on the order of ~ 1. While this is still detectable, and with the high repetition rate of the LCLS one can easily obtain spectrally dispersed data, the first experiments in both the warm dense matter regime and the plasma Thomson scattering at higher temperatures should be the independently pursued.

2. Laser Pump Probe Techniques

Since the creation of high density laser produced plasmas there have been virtually no quantitative *in situ* measurements of the kinetic rates or the populations. This is a major impediment to progress, as population kinetics of highly stripped ions is a complex problem. The complexity derives from the large number of states that must be considered in a model and the detail to which one must incorporate these states. The situation is made more difficult yet due to the fact that these plasmas tend to have rapid time evolution and large spatial gradients. [8]

Indeed, much of the effort to improve the situation has been focused on target design and advanced diagnostic development; however, the difficulties in determining the level populations or the kinetic rates remains. Therefore the interest, which comes from all areas involved with dense plasma studies and its underlying theoretical problems, *e.g.*, laboratory x-ray laser generation, laser plasma production, astrophysics, and inertial fusion, has never been met with substantial improvements in experiments.

The import of the 4th generation sources for high energy density plasma experiments is that one can use these x-ray sources to pump individual transitions in a plasma creating enhanced population in the excited states that can be easily monitored. The idea has been used in lower density plasmas with visible lasers and can, with the LCLS, be employed to advance the study of high density plasmas. [30]

Variations on the idea of pumping individual transition in high energy density plasma include the selective pumping of the wings of a line transition to observe redistribution within the line profile and pumping of selected transitions to attempt to understand the inversion mechanisms for the production of laboratory x-ray lasers. In all of these applications the tests of the theoretical developments in the areas of atomic processes, kinetics model creation, line shape formation, and x-ray gain studies would be the first of their kind as there are currently no available probes.

There are several constraints on the x-ray source for it to be useful as a laser probe of the high energy density regime. First, the probe must be tunable and this is easily satisfied. Second, the line width of the pump must be such that it can pump entire line profiles and also be capable – for studies of redistribution within line profiles – of pumping parts of the line profile. Again, these conditions will be readily met. Finally, we need to have a pump that can move enough population from one state to another so that the population changes can be monitored. This last requirement can be verified by looking at the radiative pumping rate, R_{LU}, due to the source compared to the spontaneous emission rate, A_{UL}, of the transition being pumped. This is proportional to the number of photons per mode and is given by [31]

$$\frac{R_{LU}}{A_{UL}} = 6.67x10^{-22} \frac{g_U}{g_L} \lambda_{\AA}^5 I_o^{laser} (\frac{W}{cm^2}) \frac{[,]}{\delta_\lambda \Delta_\lambda} \qquad (2)$$

where the g's are the statistical weights of the upper and lower states, λ_L and I_o are the source wavelength and intensity. The δ_λ and Δ_λ are the bandwidths of the x-ray source and the line shape of the transition being pumped, respectively, while [,] represents the minimum of the two. Two important insights emerge when evaluating equation 2. First, if we conservatively assume $I_o \sim 10^{14}$ and $[,]/\delta_\lambda \Delta_\lambda \sim 0.001$ we find that the ratio is approximately 1 for λ_L of 10 Å. This number is at least 10^3 larger than can be obtained by using a plasma source to pump a transition. Second, the ratio does not increase with decreasing source wavelength, indicating that large numbers of photons per mode will not be available as we move toward shorter wavelengths. This is due to the fact that the spontaneous rate has a strong inverse dependence on wavelength. Of course, matching, or at least controlling the source bandwidth can have salutary effects as indicated by equation 2.

The possibilities provided by plasma spectroscopic probing are illustrated with the simulation of an aluminum layer tamped on both sides with a thin layer of CH plastic irradiated by x-rays. [32] An undiluted radiation field emitted from the rear side of a 1000 Å Au target impinges on the Al foil, heating it uniformly. The Au foil is irradiated by a single 1 ns, temporally square-shaped pulse of 0.52 μm light at an intensity of $1.6x10^{14}$ W/cm^2.[33]

The emission spectrum will be the observable and it is shown in figure 5 at two times in the evolution of the plasma, one near the initiation of the x-ray pulse and the other at 18 ps later. The most notable feature is that there is a substantial increase in the hydrogenic transitions. For example, the Lyman α line at ~ 1724 eV which is unobservable in the He-like background emission without the LCLS pump, rises well above the background with the x-ray pump. Further the structure of the He-like resonance series starting at ~1598 eV and ending at the bound-free continuum near

2086 eV is substantially changed by the pumping. Indeed the Li-like satellite transitions seen on the low energy side of the He-like $1s^2$-$1s2l$ transitions are substantially enhanced. The major effect of the pump, although it is tuned to a particular transition, is to cause photoionization due to the pump strength. The ionization of the Li-like stage and the pumping of population from the He-like ground state up to the H-like ion stage cause a slow recombination decay back towards the He-like ground state.

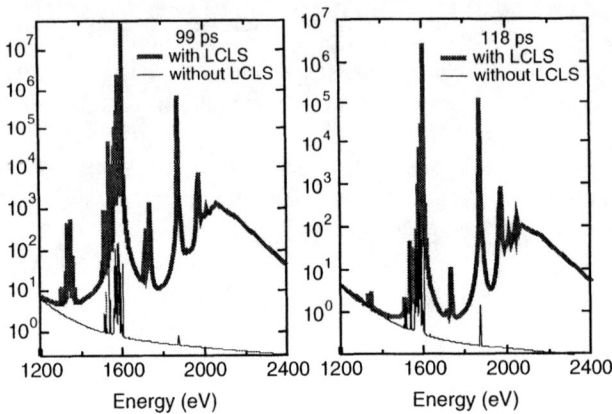

Figure 5. The logarithm of the emissivity versus spectral energy for two times in the evolution of an exploding aluminum foil. The emission from the plasma with no LCLS pump is the thin line while the thick line spectra indicate the emission when the LCLS pumps the He-like n =1 to n = 3 transition. The He-like emission, from, e.g., n =2, 3 and 4 at 1598 eV, 1868 eV, and 1963 eV, respectively, increases substantially while the H-like n=2 emission at 1724 eV arises with the pump.

It is clear from the emission spectra shown in figure 5 that one could use a fraction of the LCLS pump to generate observable signals. The detailed information that can be obtained from these measurements would provide unique constraints on the complex processes necessary to construct a complete kinetics model for the highly charged ions. Indeed, we chose to use as an example the K-shell spectra as it is easily interpretable; however, the generation of L-shell and M-shell models are also of importance and raise the level of complexity substantially. Thus, one can understand the need for experiments that can provide basic information on the processes necessary to build kinetics models.

III. SUMMARY

The next generation of light sources will bring the convergence of two experimental capabilities. On the one hand, we will have a light source with its high repetition rate and ability to serve a large user community, and on the other we will have a laser with its high peak brightness, coherence, and extremely short pulse duration. The confluence of these two capabilities will open the door for an extremely important set of experiments in the plasma and warm dense matter regimes. The discussion above has briefly touched on some of the preliminary aspects of these experiments. We end by pointing out that much more research is required.

REFERENCES

1. See, *e.g.*, K. S. Trainor, J. of Appl. Phys., **54**, 2372 (1983)
2. R. M. More *et al.*,. Phys. of Fluids, **31**, 3059 (1988)
3. See, *e.g.*, K. S. Trainor, J. of Appl. Phys., **54**, 2372 (1983)
4. R. M. More *et al.*,. Phys. of Fluids, **31**, 3059 (1988)
5. T. Guillot, Science **286**, 72 (1999)
6. L. B. Da Silva *et al.*, Phys. Rev. Lett. **78**, 483 (1997); G. W. Collins *et al.*, Science **281**, 1178 (1998)
7. T. R. Dittrich *at al.*, *Phys. Plasmas* **6**, 2164 (1999)
8. For general references to the concepts of laser-produced plasmas and plasma physics see the *Handbook of Plasma Physics*, **3**, "Physics of Laser Plasmas" edited by M. N. Rosenbluth and R. Z. Sagdeev (Elsevier, Amsterdam, 1991)
9. H.-J. Kunze, in *Plasma Diagnostics*, edited by W. Lochte-Holtgreven (North-Holland, Amsterdam, 1968)
10. J. Sheffield, *Plasma Scattering of Electromagnetic Radiation* (Academic, New York, 1975) and references therein
11. C. A. Back *et al.*, Phys. Rev. Letts. **63**, 1471 (1989); Phys. Rev. A **44**, 6730 (1991); and J. Koch *et al.*, Appl. Phys. B **58**, 7 (1994). Note that to date in higher density plasmas only the total emitted fluorescence has been studied in photopumping experiments of ion emitters.
12. for example see A. Bar-Shalom *et al.*, Phys. Rev. A **38**, 1773 (1988) and K. B. Fournier *et al.*, JQSRT **65**, 231 (2000)
13. S. H. Glenzer *et al.*, Phys. Rev. Lett. **82**, 97 (1999)
14. See the *X-Ray Lasers* 1990.edited by G. Tallents.(IOP, Bristol,1990) and AIP Proceedings X-ray Lasers, **49** (1994) and the *X-ray Lasers*, Conference series #151, 122 (IOP Publishing, Bristol) 1996.for general papers
15. See the papers in *Strongly Coupled Coulomb Systems* edited by G.J. Kalman (Plenum, New York, 1998) and the references therein
16. H. M. Van Horn, Science **252**, 384 (1991) for information on effects at 1 Mbar for hydrogen-bearing astrophysical objects
17. R. Smoluchowski, Nature **215**, 691 (1967) W. B. Hubbard, Science **214**, 145 (1981); G. Chabrier, D. Saumon, W. B. Hubbard, J. I. Lunine, Ap. J. **391**, 817 (1992); W. J. Nellis, M. Ross, N. C. Holmes, Science **269**, 1249 (1995) for Jovian planets
18. D. Saumon, et al., Ap. J. **460**, 993 (1996) for extrasolar giant planets
19. D. Saumon, W. B. Hubbard, G. Chabrier, H. M. Van Horn, A. J **391**, 827 (1992); W. B. Hubbard, et al., Phys. Plasmas **4**, 2011 (1997) for brown dwarfs
20. G. Chabrier and I. Baraffe, Astron. Astrophys. **327**, 1039 (1997) for low-mass stars
21. F. J. Rogers and D. A. Young, Phys. Rev. E **56**, 5876 (1997) and the references therein
22. The field of laser induced fluorescence is currently limited to neutral or near-neutral species. For recent examples see: D. Voslamber, Rev. Sci. Inst. **71**, 2334 (2000); J. Amorim *et al.*, J. Phys. D **33**, R51 (2000); R. E. Neuhauser *et al.*, Rev. Sci. Inst. **70**, 3519 (1999)
23. D.A. Liberman, Phys. Rev. B 20, 4981 (1979)
24. F. J. Rogers, Phys. of Plasmas **7**, 51 (2000)
25. D.A. Liberman, Phys. Rev. B 20, 4981 (1979)
26. J. Dunn, Y. Li, A. L. Osterheld, J. Nilsen, *et al.*, Phys. Rev. Lett. **84**, 4834 (2000)
27. S. Sebban, H. Daido, N. Sakaya, *et al.*, Phys. Rev. A **61**, 3810 (2000)
28. H.A. Baldis, J. Dunn, M. E. Foord, *et al.* Comments on Modern Physics, (to be published, 2000

29. The controversy on the definition of free versus bound electrons is closely connected with the role of ionization potential depression, continuum lowering, and level shifts near the continuum see the discussions in ref. [24]

30. See as examples: D. D. Burgess and C. H. Skinner, J. Phys. B **7**, L297 (1974) and G. T. Razdobarin *et al.*, Nuc. Fusion **19**, 1439 (1979)

31. R. C. Elton, *X-ray Lasers (Academic Press, San Diego, 1990)*

32. H. A. Scott and R W. Mayle, Applied Phys. B **59**, 35 (1994) and can be obtained from the website http://www.llnl.gov/def_sci/cretin

33. D. Kania *et al.* Phys. Rev. A **46**, 7853 (1992); J. Koch *et al.*, JQSRT **54**, 227 (1995); J. Moreno *et al.*, Phys. Rev. E **51**, 4897 (1995)

34. C. Mossé *et al.,* JQSRT **58**, 803 (1997) and Phys. Rev. A **60**,1005 (1999)

35. S. H. Glenzer *et al.*, JQSRT **65**, 253 (2000) and Phys. Rev E **62,** (2000)

36. See, for example, P. Volfheyn *et al.*, Phys. of Plasma **6**, 2269 (1999)

Effects of Channel Roughness on Beam Energy Spread[1]

Alexey V. Agafonov and Andrey N. Lebedev

P.N.Lebedev Physical Institute of RAS, Leninsky pr. 53, 117924 Moscow Russia

Abstract. The energy spread of particles inside a short bunch propagating along the co-axial channel with small random axially-symmetric wall perturbations is calculated.

INTRODUCTION

Short-wave free electron lasers are very sensitive to transverse and longitudinal emittances of an electron beam. The problem of radiation of electron bunches inside a channel with small random perturbations of the wall has recently been discussed. Although the radiation is small, its reaction on particles distributed along a bunch creates a spread of particle energy and leads to an uncontrolled growth of transverse emittance in the transported beam. Some estimations [1–3] of the effect show its possible danger for practically reached uniformity of the wall surface.

These estimations are based on the concept of effective beam-channel coupling impedance of a beam and a channel. Real part of the impedance is interpreted as the losses due to diffraction radiation, and imaginary part is calculated using Kronig-Kramers relations (see, for example, [4]). Radiation losses themselves are estimated for a statistical model of the surface. It seems, in our opinion, that this approach is not quite correct, or, at any case, has a vague area of applicability. Actually, the use of coupling impedance as a coefficient of proportionality between a harmonic of the current of frequency ω and of wave number k and a harmonic of electric field with the same characteristics does mean the equality of phase velocities of these two wave processes. Then, the field acting on a test particle of the monoenergetic beam depends only on the position of the particle relative to the bunch and does not depend explicitly on time. Consequently, it results in linear growth of energy deflexion and transverse pulse with the distance z.

This approach is adequate undoubtedly for systems which are uniform in the direction of propagation z. If phase velocities of proper modes of the channel exceed the velocity of the beam βc then the field is represented by Coulomb's field

---

[1] Work supported by RFFI under grant 01-02-17519.

CP581, *Physics of, and Science with, the X-Ray Free-Electron Laser*, edited by S. Chattopadhyay et al.

distorted by conducting walls and by inductive field executed negative work over the beam if the conductivity of walls is finite. There are no free proper modes (radiation) in the case. In a channel filled partially or completely with a dielectric phase velocity of some modes can be equal to the beam velocity; they create wake-field (Cherenkov's radiation) asymmetrically with respect to the bunch.

These representation with some reserve are fair also for systems with periodic perturbations, even if they are small. Due to multiple coherent reflections of waves from perturbations self-waves of a periodic system contain slow harmonics which are in synchronism with the beam and form the wake-field. However, inevitable admixture of non-resonant harmonics of self-modes results in time dependence of wake-field acting on the particle and in oscillations of particle energy accompanying systematical growth of the energy (linear resonant accelerator) or its decrease (Cherenkov's radiation). Strictly speaking, the decription in terms of coupling impedance is valid if the length of the wave is greater compared with the period. In this case periodic inhomogeneous surface can really be characterized by an effective surface impedance.

Waves propagating in the system with non-regular (random) perturberations of the wall have continious spectrum with respect to k, because reflections from individual perturbations are non-coherent. It means non-stationary behaviour of the radiation field acting on the test particle even for a harmonically space-modulated beam. Hence, the variations of a particle energy are non-regular in general case. Therefore, we calculate it using the straight way of classical electrodynamics without any artificial interpretation of the power of radiation losses as a real part of an impedance satisfying Kronig-Kramers relations. For reasons mentioned above we don't see sufficient bases for it.

EVOLUTION OF FIELD ALONG NON-REGULAR SYSTEM

We consider longitudinally modulated electron beam of velocity βc and current density $j(r, z/\beta c - t)$ which is assumed to have multiplicative distribution:

$$j(r, z, t) = \beta c q(\omega) \psi(r) \exp\left[i\omega\left(z/\beta c - t\right)\right] , \tag{1}$$

propagating co-axially along a round quasi-regular waveguide (see Fig. 1). Here, $q(\omega)$ is the Fourier-amplitude of the linear charge density, $\psi(r)$ – is a normalized to unity transverse density distribution. We limit ourselves the case of axially symmetric perturbations of the wall surface. In such waveguide only TM-waves can be excited. They influence the particle energy only and don't lead to transverse displacements. There are no principial difficulties to generalize the scheme of calculations to more general case except of rather cumbersome expressions. The deviations of the real radius of the channel $r_0(z)$ from an approximating radius a of a coaxial smooth cylinder are assumed small enough (in the sence discussed further) (see Fig. 1).

Then the longitudinal and transverse electric field satisfies nonuniform wave equations (common multiplier $\exp(-i\omega t)$ is omitted):

$$\left(\frac{1}{r}\frac{\partial}{\partial r}r\frac{\partial}{\partial r} + \frac{\partial^2}{\partial z^2} + k_0^2\right)E_z = i\frac{4\pi k_0}{\beta\gamma^2}q(k_0)\psi(r)\exp(ik_0 z/\beta) ; \tag{2}$$

$$\left(\frac{\partial}{\partial r}\frac{1}{r}\frac{\partial}{\partial r}r + \frac{\partial^2}{\partial z^2} + k_0^2\right)E_r = 4\pi q(k_0)\frac{\partial\psi(r)}{\partial r}\exp(ik_0 z/\beta) , \tag{3}$$

where

$$\gamma = \left(1 - \beta^2\right)^{-1/2} ; \quad k_0 = \omega/c .$$

Instead of the second equations it is convenient to use the equation following from $\operatorname{div}\vec{E} = 4\pi\rho$:

$$E_r = \frac{4\pi q(k_0)}{r}\int_0^r \psi(r)r\,\mathrm{d}r - \frac{1}{r}\int_0^r \frac{\partial E_z}{\partial z}r\,\mathrm{d}r . \tag{4}$$

Equations (2, 3) must be completed by the following boundary condition

$$E_z\left(r_0(z), z\right) \equiv -\frac{\mathrm{d}r_0}{\mathrm{d}z}E_r\left(r_0(z), z\right) . \tag{5}$$

It means vanishing of a tangential component of the electric field on the perfectly conducting wall of the waveguide.

Following methodically [1,2] we approximate the boundary condition by its expansion at $r = a$ using the smallness of $\delta(z) = a - r_0(z)$:

$$E_z\left(a, z\right) \approx \delta' E_r\left(a, z\right) + \delta\left.\frac{\partial E_z}{\partial r}\right|_{r=a} + \dots , \tag{6}$$

where prime denotes a total derivative with respect to z. This approximation is reasonable if $\delta' \ll 1$, i.e, if perturbations at the surface are smooth enough.

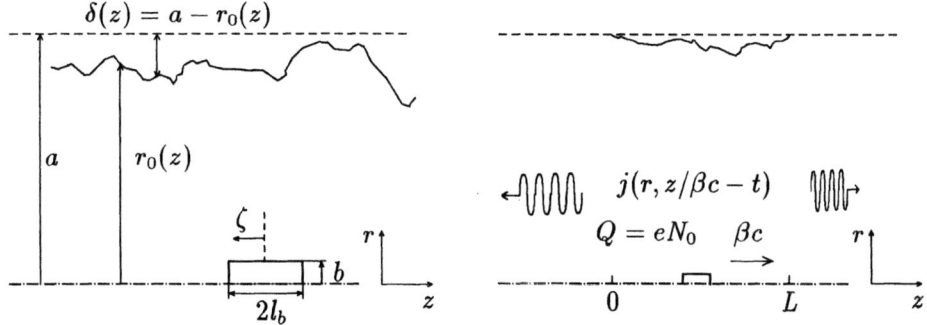

FIGURE 1. Geometry and notation.

The area of its applicability was discussed in [1,2] and can be found in details, for example, in [5]. Under condition $|\delta'| \ll 1$, when the pertuberations are smooth, the fields excited by the beam inside a regular wavegude of radius a can be substituted into the right part of (6). Because the radiation can't exist in a regular waveguide (except of transition radiation at the input and the output of the waveguide) then the zero order field propagates with the velocity of the beam having periodicity $\propto \exp{(ik_0 z/\beta)}$ and represents just a Coulomb's field in the laboratory frame. The method of its calculations is a standart one. It is necessary to use the expansion the field over the full system of eigenfunctions of the transverse Laplacian's operator (for the case, Bessel's functions of zero order with roots λ_n). Representing transverse distribution as

$$\psi(r) = \sum_n \psi_n J_0 \left(\lambda_n \frac{r}{a} \right) \tag{7}$$

and mark "zero order" fields by the superscript 0 it follows after simple transformations

$$E_z^0 = -i \sum_n \frac{4\pi k_0 a^2 \beta \psi_n q(k_0)}{\lambda_n^2 \beta^2 \gamma^2 + k_0^2 a^2} J_0 \left(\lambda_n \frac{r}{a} \right) \exp{(ik_0 z/\beta)} \; ;$$

$$\frac{\partial E_z^0}{\partial r} = i \sum_n \frac{4\pi k_0 a \lambda_n \psi_n \beta q(k_0)}{\lambda_n^2 \beta^2 \gamma^2 + k_0^2 a^2} J_1 \left(\lambda_n \frac{r}{a} \right) \exp{(ik_0 z/\beta)} \; ; \tag{8}$$

$$E_r^0 = \sum_n \frac{4\pi a \lambda_n^2 \beta^2 \gamma^2 q(k_0) \psi_n}{\lambda_n^2 \beta^2 \gamma^2 + k_0^2 a^2} J_1 \left(\lambda_n \frac{r}{a} \right) \exp{(ik_0 z/\beta)} \; .$$

Taking into account these expressions the boundary condition (6) can be written as

$$E_z(a, z) = Aq(k_0) f(z) \exp{(ik_0 z/\beta)} \; , \tag{9}$$

where

$$f(z) = \delta'(z) + ik_0 \delta/\beta \gamma^2 \; ; \qquad A = \sum_n \frac{4\pi a \lambda_n^2 \beta^2 \gamma^2 \psi_n}{\lambda_n^2 \beta^2 \gamma^2 + k_0^2 a^2} J_1 (\lambda_n) \; . \tag{10}$$

The solution of the nonuniform equation (2) for E_z with the nonuniform boundary condition (9) can be represented in the following form:

$$E_z(r, z) = E_z^0(r, z) + Aq(k_0) f(z) \exp{(ik_0 z/\beta)} + E_z^1(r, z) \tag{11}$$

where $E_z^1(r, z)$ satisfies the nonuniform equation with zero boundary condition:

$$\frac{1}{r} \frac{\partial}{\partial r} r \frac{\partial E_z^1}{\partial r} + k_0^2 E_z^1 + \frac{\partial^2}{\partial z^2} E_z^1 = -Aq(k_0) \left[f(z) \exp{(ik_0 z/\beta)} \right]'' \; . \tag{12}$$

Note once more that $E_z^0(r, z)$ and $E_z^0(a, z)$ have no poles for real k_0 and, consequently, can't describe free propagating radiation field. The last is described by the term $E_z^1(r, z)$ and have to meet the demands of the Sommerfeld's principle (principle of causality). For the case: at the input of the waveguide ($z = 0$) waves exist and propagate only in negative direction of z-axes; and at the output of the waveguide ($z = L$) waves propagate only in positive direction. This rule of the solution choice is valid and for $L \to \infty$ as well.

As far as $E_z^1(a, z) = 0$ we can find the solution of (12) in the form of the expansion over the same eigenfunctions $J_0\left(\lambda_m \frac{r}{a}\right)$:

$$E_z^1(r, z) = \sum_m E_m(z) J_0\left(\lambda_m \frac{r}{a}\right) . \tag{13}$$

Then

$$\sum_m \left[\left(k_0^2 - \frac{\lambda_m^2}{a^2}\right) E_m + E_m''\right] J_0\left(\lambda_m \frac{r}{a}\right) = -Aq(k_0) \left[f(z) \exp\left(ik_0 z/\beta\right)\right]'' . \tag{14}$$

Multiplying (14) by $rJ_0\left(\lambda_s \frac{r}{a}\right)$, integrating over r from 0 to a, and taking into account orthogonality of the eigenfunctions we obtain the equation for complex amplitudes of the radiation field:

$$\left(k_0^2 - \frac{\lambda_s^2}{a^2}\right) E_s + E_s'' = -\frac{2Aq(k_0)}{\lambda_s J_1(\lambda_s)} \left[f(z) \exp\left(ik_0 z/\beta\right)\right]'' . \tag{15}$$

The solution of this total derivative equation is

$$E_s(z) = A_s^+(z) \exp\left(i\kappa_s z\right) + A_s^- \exp\left(-i\kappa_s z\right) - \frac{2Aq(k_0)f}{\lambda_s J_1(\lambda_s)} \exp\left(ik_0 z/\beta\right) , \tag{16}$$

where

$$\kappa_s = \left(k_0^2 - \lambda_s^2/a^2\right)^{1/2} ; \qquad A_s^{+\prime}(z) \exp\left(i\kappa_s z\right) + A_s^{-\prime} \exp\left(-i\kappa_s z\right) = 0 .$$

The amplitudes of accompanying (forward) and backdirected waves are described by the following equations:

$$A_s^{+\prime}(z) = -i\frac{Aq(k_0)\kappa_s f(z)}{\lambda_s J_1(\lambda_s)} \exp\left(i\left(k_0/\beta - \kappa_s\right) z\right) ;$$

$$\tag{17}$$

$$A_s^{-\prime}(z) = i\frac{Aq(k_0)\kappa_s f(z)}{\lambda_s J_1(\lambda_s)} \exp\left(i\left(k_0/\beta + \kappa_s\right) z\right) .$$

Taking into account above mentioned principle of causality we have

$$A_s^+ = -\frac{iA\kappa_s q(k_0)}{\lambda_s J_1(\lambda_s)} \int_0^z f(x) \exp\left(i\left(k_0/\beta - \kappa_s\right)x\right) dx \, ;$$

(18)

$$A_s^- = -\frac{iA\kappa_s q(k_0)}{\lambda_s J_1(\lambda_s)} \int_z^L f(x) \exp\left(i\left(k_0/\beta + \kappa_s\right)x\right) dx \, .$$

Consequently, a spatially and temporally modulated current of large enough frequency (higher than the cutoff frequency $\lambda_s c/a$) excites forward and backdirected modes of radiation. Low frequency components of the current (imaginary κ_s) do not excite radiation, and corresponding fields propagate together with the beam. These fields represent corrections of Coulomb's field more or less locally related to the current density. Note, that the correction themselves are calculated without taking into account boundary effects at $z = 0$ and $z = L$. However this can be of importance only if $k_0 a \approx \lambda_s$, when Coulomb's field is not yet screened by the walls.

The radiation field as distinct from Coulomb's field is defined by overall beam inside the waveguide, and grow (for forward wave) along it. However, because of random positions of small pertuberations of the wall surface the phase of the field is random as well, and the radiation can be only partially coherent even a perfect regular arrangement of primary radiators i.e., of beam particles. The degree of the coherency is defined by correlation properties of a nonunoform surface.

PARTICLE ENERGY VARIATIONS DUE TO RADIATION FIELD

Due to the reasons discussed above we can consider only the variation of particle energy forced by the radiation field. Actually, the distortions of Coulomb's field due to small wall roughness have to be smal compared with the main Coulomb's field inside a smooth waveguid, by definition. At the same time the last effect is assumed *a priori* as being not dangerous. For the simplicity we don't take into account also the backdirected radiation wave which is wittingly far from the synchronism with particles.

However, even accompanying radiation field is not in exact synchronism with particles because phase velocity of the wave always exceeds c. That is why the energy of particles changes along z not monotonously but oscillates with the larger amplitude the smaller the difference of phase velocity k_0/κ_s and particle velocity β (note that these oscillations of energy in the field of a mode do not permit to use the conception of a longitudinal impedance). Moreover, the amplitude of the oscillations grows along the distance as well as the amplitude of radiation field does. It is easy to foresee that the definite influence of the accompanying wave on particles takes place if particles are ultrarelativistic and if frequences of modes are higher than the cutoff frequency (phase velocity close to c).

We find the energy variation $\Delta_s \gamma(\zeta)$ of a test particle of charge e placed at a lag distance ζ ($\zeta = 0$ at the center of a bunch) in longitudinal direction and at radial position r under the action of s-mode. We restore omitted term $\exp(-ik_0 ct)$ in the field expression putting there $t = (z + \zeta)/\beta c$, and integrate $J_0(\lambda_s r/a) E_s(z) \exp(-ik_0 z/\beta)$ along the waveguide (the influence of Coulomb's field of zero order and backdirected wave is neglected):

$$\Delta_s \gamma(\zeta) = -i \frac{Aq(k_0)e\kappa_s J_0\left(\lambda_s \frac{r}{a}\right)}{mc^2 \lambda_s J_1(\lambda_s)} \exp(-ik_0 \zeta/\beta) \times$$

$$\int_0^L \int_0^z f(x) \exp[i(\kappa_s - k_0/\beta)(z-x)] \, dx \, dz \, . \tag{19}$$

Integrating (19) by parts and taking into account expression (10) for $f(z)$ we have:

$$\Delta_s \gamma(\zeta) = -i \frac{Aq(k_0)e\kappa_s J_0\left(\lambda_s \frac{r}{a}\right)}{mc^2 \lambda_s J_1(\lambda_s)(\kappa_s - k_0/\beta)} \exp(-ik_0 \zeta/\beta) \times$$

$$\int_0^L \delta(z) \left[(\kappa_s - k_0\beta) \exp[i(\kappa_s - k_0/\beta)(L-z)] - \frac{k_0}{\beta \gamma^2}\right] dz \, . \tag{20}$$

We put here $\delta(L) = \delta(0) = 0$. This condition could be provided with smooth end regions of the waveguide shown in Fig. 1. Anyway, such details have no physical meaning because the radiation condition accepted above has a qualitative character as it neglects edge effects.

The obtained expression has a simple interpretation: a forward directed radiation wave born at point z convoys the particle along a distance $L - z$ slipping with a period $2\pi/(\kappa_s - k_0/\beta)$ and changing periodically the particle's energy.

Various harmonics of the current are not equal in this process. The minimal phase slip corresponds to the harmonic determined by

$$\frac{d(\kappa_s - k_0/\beta))}{dk_0} = 0 \, , \quad \text{or} \quad k_0 = k_s^* = \pm \frac{\lambda_s \gamma}{a} \, . \tag{21}$$

For this harmonic $\kappa_s - k_0/\beta = \mp \lambda_s/a\beta\gamma$. One can see that for a relativistic particle the slip can be really small and rather large energy oscillations can be expected if this quasiresonant harmonic and its nearest neighbours are well presented in the bunched beam spectrum.

Note that the second term in brackets in (20) is of the same type as the last "Coulomb" term omitted in (16). Both of them are proportional to $\int \delta \, dz$. Nethertheless, the radiation term contains an additional factor of order of $\kappa_s/(\kappa_s - k_0/\beta)$ which can be large for the quasiresonant mode. In general, this a common feature of radiation fields being compared with the Coulomb ones.

The larger is the distance the narrower is the frequency band of harmonics influencing effectively a particle energy. Hence, the expression (19) should be integrated over k_0, i.e., the inverse Fourier transformation is to be performed to restore a real spatial/temporal structure of the bunched beam field.

Being integrated over k_0 the expression (20) yields:

$$\Delta_s\gamma = \frac{ieJ_0\left(\lambda_s\frac{r}{a}\right)}{mc^2\lambda_s J_1\left(\lambda_s\right)} \int_0^L \delta(L-\xi)G(\xi)\,\mathrm{d}\xi\,, \tag{22}$$

where

$$G(\xi) = -\int_{-\infty}^{+\infty} Aq(k_0)\kappa_s \exp\left(-ik_0\zeta/\beta\right) \times$$
$$\left\{1 + \frac{\kappa_s - k_0\beta}{\kappa_s - k_0/\beta}\Big[\exp\left[i(\kappa_s - k_0/\beta)\xi\right] - 1\Big]\right\}\,\mathrm{d}k_0\,. \tag{23}$$

According to general rules, the integration over k_0 should be performed in the upper complex half-plane which corresponds to the field absence at $t \to -\infty$. Integrating along the real axis one should put $k_0 = \mathrm{Re}\,k_0 + i0$ and

$$\kappa_s^2 = \mathrm{Re}^2 k_0 + 2i0 \times \mathrm{Re}\,k_0 - \lambda_s^2/a^2.$$

But for the forward directed wave the real part $\mathrm{Re}\,\kappa_s$ must have the same sign as $\mathrm{Re}\,k_0$ has. Thus,

$$\kappa_s = -\sqrt{k_0^2 - \lambda_s^2/a^2} \quad \text{for} \quad \mathrm{Re}\,k_0 < -\lambda_s/a\,;$$

$$\kappa_s = +\sqrt{k_0^2 - \lambda_s^2/a^2} \quad \text{for} \quad \mathrm{Re}\,k_0 > \lambda_s/a\,.$$

For $|\mathrm{Re}\,k_0| < \lambda_s/a$ always $\kappa_s = i\sqrt{\lambda_s/a - k_0^2}$.

We shall suppose in what follows that $q(k_0)$ is an even function, which is equivalent to a longitudinal symmetry of the bunch. Bearing in mind that $A(k_0)$ is an even function as well and using the above choice of κ_s one can integrate over the real variable k_0 which is more convenient for computing. For the analysis it is more convinient to repesent $G(\xi)$ in the form of the sum of two functions, $G(\xi) = I_1 - I_2(\xi)$, where

$$I_1 = -\frac{2i}{\beta\gamma^2} \int_{\lambda_s/a}^{\infty} \frac{Ak_0 q(k_0)\kappa_s}{\kappa_s - k_0/\beta} \sin\frac{k_0}{\beta}\zeta\,\mathrm{d}k_0 -$$

$$\tag{24}$$

$$-\frac{2i}{\beta\gamma^2} \int_0^{\lambda_s/a} \frac{Ak_0 q(k_0)|\kappa_s|}{\lambda_s^2/a^2 + k_0^2/\beta^2\gamma^2} \left(|\kappa_s|\sin\frac{k_0}{\beta}\zeta + \frac{k_0}{\beta}\cos\frac{k_0}{\beta}\zeta\right)\mathrm{d}k_0\,;$$

$$I_2 = -2i \int_{\lambda_s/a}^{\infty} \frac{Aq(k_0)\kappa_s\left(\kappa_s - k_0\beta\right)}{\kappa_s - k_0/\beta} \sin\left[\frac{k_0}{\beta}\zeta - \left(\kappa_s - \frac{k_0}{\beta}\right)\xi\right]\mathrm{d}k_0 -$$

$$-2\mathrm{i}\int_0^{\lambda_s/a}\frac{Aq(k_0)|\kappa_s|\exp\left(-|\kappa_s|\xi\right)}{\lambda_s^2/a^2+k_0^2/\beta^2\gamma^2}\times$$

$$\times\left[\frac{k_0|\kappa_s|}{\beta\gamma^2}\sin\frac{k_0}{\beta}\left(\zeta+\xi\right)-\frac{\lambda_s^2}{a^2}\cos\frac{k_0}{\beta}\left(\zeta+\xi\right)\right]\mathrm{d}k_0\,.$$

Bearing in mind that

$$\frac{\mathrm{d}}{\mathrm{d}k_0}\left(\kappa_s-\frac{k_0}{\beta}\right)=\frac{\beta k_0-\kappa_s}{\beta\kappa_s}\,,$$

and integrating by parts the full expression for I_2 can be also reduced to

$$I_2=-\frac{\mathrm{i}\beta}{\xi}\int_{-\infty}^{\infty}\exp\left[\mathrm{i}\left(\kappa_s-k_0/\beta\right)\xi\right]\frac{\mathrm{d}}{\mathrm{d}k_0}\frac{Aq(k_0)\kappa_s^2}{\kappa_s-k_0/\beta}\exp\left(-\mathrm{i}k_0\zeta/\beta\right)\mathrm{d}k_0\,. \qquad (26)$$

The behaviour of $G(\xi)$ depends on the relative values of I_1 and I_2 and is determined by an ierarchy of three characteristic lengths – the channel transverse radius a, the channel length L (or the distance $\xi=L-z$), and the bunch longitudinal size in the lab frame l_b. Apart from pure geometry this ierarchy depends essentially on the Lorentz factor γ.

To make the latter clear let us cosider a phase slip per unit length of the radiation field relative to a particle. As was mentioned above it is minimal at $k_0=\lambda_s\gamma/a$ being equal to $-\lambda_s/a\beta\gamma$. However, this minimum is rather smooth in a relativistic case, i.e. the slip is nonsensitive to the harmonic number within the region $1\ll k_0a/\lambda_s\ll\gamma^2$ because all short-wave (wavelength smaller than a) harmonics propagate with velocity practically equal to that of light. Hence, two limiting cases are to be distinguished.

1. If the independent variable ξ satisfies the inequality

$$\frac{a\gamma}{\lambda_s\xi}\gg 1\,, \qquad (27)$$

then for the majority of the harmonics (with wavenumbers essentially lesser than $\lambda_s\gamma^2/a$) the phase slip at the total length is small. The corresponding exponent in (23) may be changed then for unity. Then $G(\xi)\approx G(0)$ is independent of ξ and the expression for the enegy gain takes the simple form

$$\Delta_s\gamma\approx\frac{\mathrm{i}eJ_0\left(\lambda_s\frac{r}{a}\right)}{mc^2\lambda_sJ_1\left(\lambda_s\right)}G(0)\int_0^L\delta(L-\xi)\,\mathrm{d}\xi\,, \qquad (28)$$

where

$$G(0)=-\int_{-\infty}^{+\infty}Aq(k_0)\kappa_s\exp\left[-\mathrm{i}k_0\zeta/\beta\right]\mathrm{d}k_0=$$

$$(29)$$

$$=2\mathrm{i}\int_{\lambda_s/a}^{\infty}Aq\kappa_s\sin(k_0\zeta/\beta)\,\mathrm{d}k_0-2\mathrm{i}\int_0^{\lambda_s/a}Aq\,|\kappa_s|\cos(k_0\zeta/\beta)\,\mathrm{d}k_0\,.$$

This case can be considered as the "short channel" solution if the length of the channel L satisfies (27) or:

$$\frac{a\gamma}{\lambda_s} = L^* \gg L . \tag{30}$$

The physical reason for this insensibility to the perturbation local structure is obvious. If the condition (27) is fulfilled the channel, as a whole, looks in the particle rest frame like a short ring with longitudinal dimension much smaller than its radius. Hence, the radiation under consideration should be treated as an addition to the transition radiation due to edge effects.

In the opposite case it is necessary to divide the integral (20) at least on in parts:

$$\int_0^L \delta(L-\xi)G(\xi)\,d\xi = \int_0^{L^*} \delta(L-\xi)G(0)\,d\xi + \int_{L^*}^L \delta(L-\xi)G(\xi)\,d\xi . \tag{31}$$

Of course, we ignore here the transition region near $\xi = L^*$ if $L \gg L^*$ and as it follows the exression (31) is considered as an estimation.

2. If $a\gamma/\lambda_s = L^* \ll \xi$ even a smooth minimum of the function $\kappa_s - k_0/\beta$ plays certain selective role and the quasiresonant harmonic can be of importance. The radiation field is more structurized then and may ifluence the particle energy more effectively not being smeared by other harmonics. However this selected role of the harmonic with the minimal phase slip (21) can be realized if the harmonic is well presented in a $q(k_0)$ distribution, i.e., if the bunch is short enough. Then the integral I_2 can be readily evaluated using the stationary phase method (see, for example, [6]). The argument of the oscillating function in (25) containing a large parameter ξ can be presented in the vicinity of the stationary phase point $k^* = \lambda_s\gamma/a$ as

$$\xi\left(\kappa_s - \frac{k_0}{\beta}\right) \approx \xi\left(-\frac{k^*}{\beta\gamma^2} - \frac{\lambda_s^2}{2a^2 k^{*3}\beta^3}(k_0 - k^*)^2 + ...\right) . \tag{32}$$

After some simple arithmetic one gets

$$I_2 \approx -\frac{2\beta^2\gamma a}{\lambda_s\xi}\sqrt{\frac{2\pi\kappa^{*3}}{\xi}}\left.\frac{d}{dk_0}\right|_{k_0=k^*} Aq\kappa^2 \sin\left(\frac{\xi}{\beta\gamma} + \frac{k_0\zeta}{\beta} + \frac{\pi}{4}\right) . \tag{33}$$

Note that the combination $\kappa_s - k_0/\beta$ is not differentiated and is put equal to $\pm 1/\gamma\beta$ because its derivative is zero at the stationary phase point. The derivative with respect to k_0 is taken at the point $k_0 = \lambda_s\gamma/a$.

The expression (33) diverges at $\xi \to 0$ and is asymptotic being valid for large ξ only. We make this evaluation bearing in mind two aims. Firstly, if the bunch is not too short $I_2 \ll I_1$ at the most part of the channel, so that $G(\infty) \approx I_1$. Then

$$\Delta_s\gamma \approx \frac{ieJ_0\left(\lambda_s\frac{r}{a}\right)}{mc^2\lambda_s J_1(\lambda_s)}G(\infty)\int_{L^*}^L \delta(L-\xi)\,d\xi . \tag{34}$$

Secondly, I_2 oscillates along z with rather unexpectedly small wavenumber $\lambda_s/a\beta\gamma$. Hence, it selects quite low frequency harmonics from the perturbation spectrum, if at all.

Taking into account expression (31) the energy gain for the case $\xi \gg L^*$ can be approximately written as

$$\Delta_s\gamma = \frac{ieJ_0\left(\lambda_s\frac{r}{a}\right)}{mc^2\lambda_s J_1\left(\lambda_s\right)}\left[G(0)\int_0^{L^*}\delta(L-\xi)\,\mathrm{d}\xi + G(\infty)\int_{L^*}^L\delta(L-\xi)\,\mathrm{d}\xi\right]. \qquad (35)$$

Gaussian bunch. It follows from the consideration above that the total intensity as well as the amplitude of the most effective harmonic depend on a spatial distribution of the bunch current. We shall restrict ourselves by the case of a transersally uniform $(0 < r < b)$ single bunch of charge Q with Gaussian longitudinal distribution of halfwidth l_b:

$$j_0(r,z,t) = \frac{c\beta Q}{\pi\sqrt{2\pi}b^2 l_b}\exp\left[-\frac{(z-\beta ct)^2}{2l_b^2}\right]; \qquad r < b. \qquad (36)$$

Temporal harmonics are then determined as

$$q(k_0) = \frac{cQ}{(2\pi)^{3/2}l_b}\int_{-\infty}^{+\infty}\exp\left[-\frac{(z-\beta ct)^2}{2l_b^2} - i\frac{k_0}{\beta}(z-\beta ct)\right]\,\mathrm{d}t = \qquad (37)$$

$$= \frac{Q}{2\pi}\exp\left(-k_0^2 l_b^2/2\beta^2\right).$$

Transverse spatial harmonics follow from Bessel transformation:

$$\psi_n = \frac{2J_1\left(\lambda_n b/a\right)}{ab\lambda_n J_1^2\left(\lambda_n\right)}. \qquad (38)$$

Then

$$A(k_0) = \sum_n \frac{8\pi\lambda_s\beta^2\gamma^2}{b\left(\lambda_n^2\beta^2\gamma^2 + k_0^2 a^2\right)}\frac{J_1\left(\lambda_n b/a\right)}{J_1\left(\lambda_n\right)}. \qquad (39)$$

RANDOM PERTURBATIONS

The estimates (27) and (34) obtained above are valid for an arbitrary realization of the perturbation $\delta(z)$ and contain certain information on the asymptotic phase of the radiation field. In the case of random perturbation this information, of course, is not of importance. What really counts is a slow dependence of the amplitude of $I_2(z)$ on the longitudinal coordinate due to field accumulation along the channel.

The value of $I_2(z)$ also oscillates slowly because of the phase slip which is finite even for the quasiresonant harmonic.

Applicability of the esimates above needs several remarks. The stationary phase method implays existence of a quasiresonant harmonic determined by a selective function. The role of that function is played in our case by

$$\exp\left[-i\frac{\xi\lambda_s^2}{2a^2 k^{*3}\beta^3}(k_0 - k^*)^2\right] = \exp\left[\pm i\frac{\xi a}{2\lambda_s\gamma^3\beta^3}(k_0 - k^*)^2\right]. \tag{40}$$

The width of the function along k_0 axis is equal by order of magnitude to $(\lambda_s\gamma^3/a\xi)^{1/2}$ and decreases for $\xi \to \infty$ Actually, this is this feature that determines the asymptotic behaviour of $I_2(\xi)$. To apply the stationary phase method the function $\propto \exp(-l_b^2 k_0^2/2\beta^2)$ has to be almost constant within this interval. However it can be sharply selective as well if the bunch length l_b is large enough. The unequality

$$l_b \ll \left(\frac{a\xi}{\gamma^3\lambda_s}\right)^{1/2} \tag{41}$$

necessary for applicability of the estimations can easily be violated even for a long channel $(\xi \gg a\gamma)$ where the phase slip of the field is large. So, the foreseen behaviour

$$I_2 \propto \exp\left(-\frac{l_b^2\gamma^2\lambda_s^2}{2\beta^2 a^2}\right)$$

for $l_b \to \infty$ overestimates the effect and is valid only if (41) is fulfilled. For larger values of l_b the functions $\exp(-l_b^2 k_0^2/2\beta^2)$ and (40) can not overlap at all. By the way, even more conservative estimate

$$I_2 \propto \exp(-\xi\lambda_s/a\gamma) \qquad \text{for} \quad l_b \to \infty$$

for a long channel limits the energy spread at a quite low level.

Bearing that in mind we may suppose that $I_2(\xi)$ does not change at the perturbations characteristic length. In our case of random perturbations the latter has to be equal to the correlation length l_c.

Now we can come back to the general expression (22), supposing that the mathematical expectation $\bar{\delta} = 0$. As far as I_2 is a slow function of z and does not change its value at the correlation length l_c the mathematical expectation $\bar{\Delta}_s$ vanishes as well. The role of a real characteristic of the energy diffusion is played by the diffusion coefficient $\langle\Delta^2\rangle$ which is a sum over transverse harmonics squared and averaged:

$$\langle\Delta^2\rangle = \left\langle\left|\sum_s \Delta_s\gamma\right|^2\right\rangle$$

Note that according to (20) this expression can be written as

$$\langle \Delta^2 \rangle = \int_0^L \int_0^L \langle \delta(L - \xi)\delta(L - \xi')G(\xi)G(\xi') \, d\xi \, d\xi' . \tag{42}$$

If one neglects edge effects $\delta(L - \xi)$ represents a stationary random process with a correlation function

$$F(|\xi - \xi'|) = \langle \delta(L - \xi)\delta(L - \xi') \rangle$$

which depends on $|\xi - \xi'|$ only and vanishes for $|\xi - \xi'| \gg l_c$. Defining the correlation length as

$$l_c = \frac{\int_0^\infty F(x) \, dx}{\langle \delta^2 \rangle}$$

and bearing in mind that $G(\xi)$ is almost constant at this length one obtains

$$\langle \Delta^2 \rangle = \langle \delta^2 \rangle l_c \int_0^L |G(\xi)|^2 \, d\xi . \tag{43}$$

Note that for $|G| \approx$ Constant $\langle \Delta^2 \rangle \propto l_c L$ as it should be for a diffusion process.

RESULTS AND ESTIMATIONS

From expressions (35) and (43) it follows that the energy gain at the output of the channel can be written as

$$\langle \Delta^2 \rangle = C^2 G^{*2} \langle \delta^2 \rangle l_c L ,$$

where $C = r_0 N_0 / ba^2$, r_0 – is the electron classical radius, and $G^* = G^*(0)$ or $G^* = G^*(\infty)$ are dimensionless functions (see Fig. 2) of the lag ζ (measured in channel radius a units) describing the distribution of energy gain along the bunch.

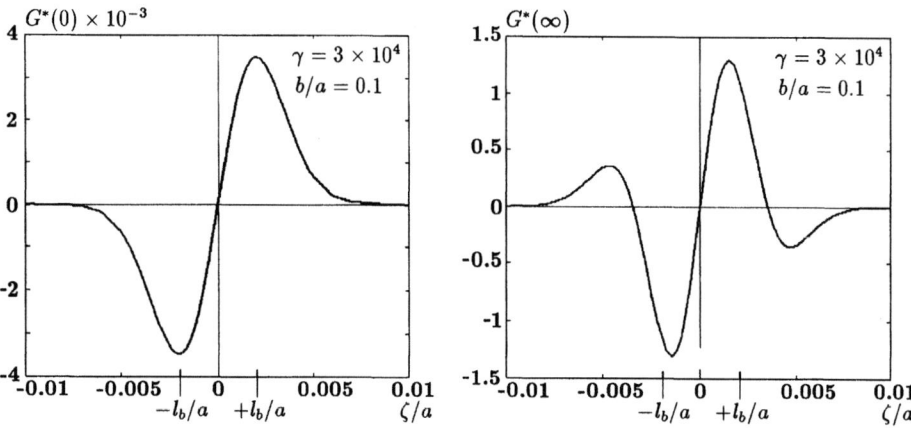

FIGURE 2. Dimensionless part of particle energy variations. The dependences are slightly assymetric with respect to the center of the bunch.

Calculations were carried out for the set of parameters of LCLS: $\gamma = 30000$; $l_b/a = 0.002$ (for radius of the channel $a = 0.5$ cm the half-length of the bunch is $l_b = 10\,\mu m$). The coefficient $C = 1.4 \times 10^3\,m^{-2}$ for $b/a = 0.1$ and bunch charge $Q = 1$ nC ($N_0 = 6.25 \times 10^9$). For the set of the parameters it follows from (30) that $L^* \approx 50$ m. Taking maximum values of dimensionless energy gains from Fig. 2 and choosing ruther pessimistic parameters $\sqrt{\langle \delta^2 \rangle} \approx 5\,\mu m$, $l_c = a$ and $L \gg L^*$ we find the amplitude of energy oscillations $\Delta\gamma/\gamma \leq 5 \times 10^{-4}$.

We consider the energy spread inside a bunch propagating along the channel with small random perturbations of the wall as not dangerous for the X-ray FEL's. From this point of view a more vital problem is the energy spread forced by self space charge of the bunch.

REFERENCES

1. Stupakov G.V., *Phys. Rev. Special Topics – Accelerators and Beams.* **1**, 064401 (1998).
2. Stupakov G., Thompson R.E., Walz D., Carr R., *Phys. Rev. Special Topics – Accelerators and Beams.* **2**, 064401 (1999).
3. Novokhatski A., Mosnier A., *Proc. of the 1997 Particle Accelerator Conference*, Vancouver, B.C. **2**, 1661 (1997).
4. Landau, L., and Lifshits, E., *Electrodynamics of Continuous Media*, Moscow: Nauka, 1982.
5. Rytov, S.M., *Vvedenie v Statisticheskuyu Radiofisiku (Introduction to Statistical Radiophysics)*, Moscow: Nauka, 1966.
6. De Bruijn, N.G., *Asymptotic Methods in Analysis*, North Holl. Publ. Co., 1958.

Field characteristics of novel hybrid undulators adequate to LCLS requirements. Simulation results.

A.A.Varfolomeev, S.V.Tolmachev

Coherent Radiation Laboratory, RRC "Kurchatov Institute", 123182 Moscow

Abstract. Magnetic field characteristics of two novel hybrid undulator schemes are given. The undulators were specially designed to be adequate to the X-ray FEL project LCLS. It is demonstrated that using these novel schemes the high specified magnetic field strengths can be obtained at larger gaps than with any other known hybrid undulator scheme. Basic field parameters are given for the planar and the helical hybrid undulators with the specified period 3 cm.

INTRODUCTION

Large X-ray FEL projects like LCLS [1] require very specific undulators. Strong magnetic fields at not small gaps, great whole length, high accuracy for the field strengths and construction alignment through the entire undulator length are typical demands for these projects. Due to long undulator lengths and required small electron beam emittance and energy spread these FEL performances become more sensitive to vacuum pipe diameter limited by the undulator gap. To decrease wake field effect caused by resistivity and roughness of the beam pipe walls larger vacuum pipe diameters are needed [2]. So the requirements look as contradictory: high field strength (possible at small gaps) from one side and not low limit for the gap (providing small enough wake field effects) from other side.

Present work was done with the purpose to find new solution for the undulator scheme and construction that would be better for the LCLS project than the ordinary undulator construction being used in the up-date LCLS project design [1].

NOVEL UNDULATOR SCHEMES AND FIELD SIMULATION RESULTS

The following basic parameters were used as required for fitting the main goal of

CP581, *Physics of, and Science with, the X-Ray Free-Electron Laser*, edited by S. Chattopadhyay et al.

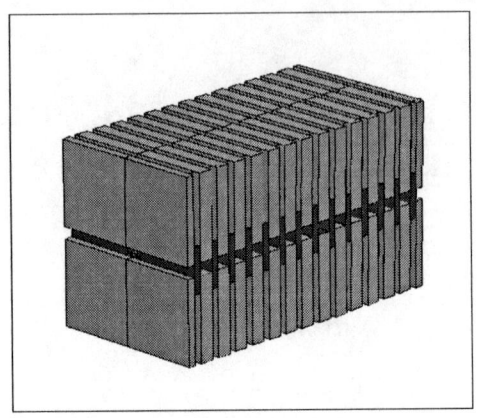

FIGURE 1. Schematic view of the planar hybrid undulator construction KIAE-3P.

the project [1]:

Field amplitude of a planar undulator construction	$B_o = 1.32$ T;
Field amplitude of a helical undulator construction	$B_o = 0.93$ T;
Undulator period	$\lambda_w = 30$ mm;
Undulator gap	$g \geq 6$ mm.

No one known before scheme could provide such fields. It was decided to develop both planar and helical hybrid undulator schemes with the purpose to obtain field strengths not lower than above nominal field strengths at gaps exceeding 6 mm as much as possible. New hybrid schemes were analyzed with using RADIA code. Space room was used more efficiently for permanent magnets insertion, some new kinds of side magnets were included, optimization of dimensions and shapes of permanent magnet and Va permendur components was done. As a result it was found indeed that the above specified fields can be provided for larger than 6 mm gaps. These results were achieved for both the planar and the helical undulator configurations.

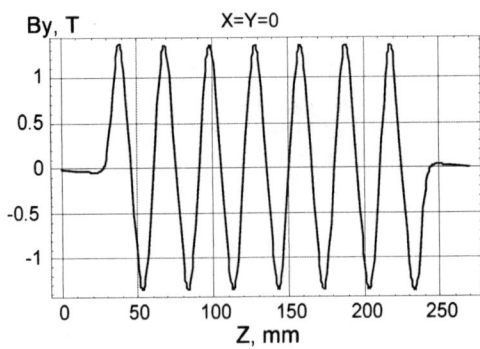

FIGURE 2. Transverse magnetic field B_y in the KIAE-3P undulator model as a function of along the undulator axis distance. Undulator period $\lambda_w = 30$ mm, gap $g = 8.5$ mm, on-axis peak field $B_o = 1.36$ T.

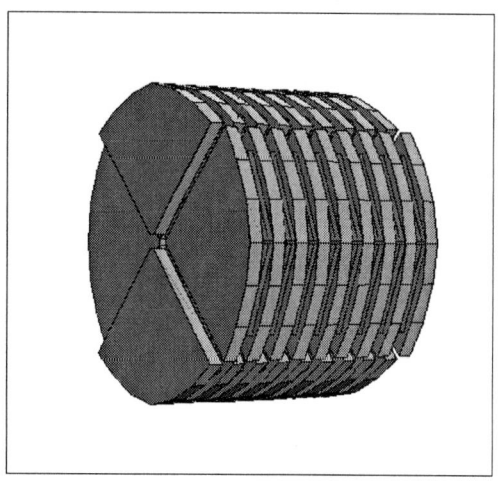

FIGURE 3. Schematic view of the helical hybrid undulator model KIAE-3H.

Schematic view of the optimized planar hybrid undulator construction is shown in Fig. 1. It is seen that all mechanical components are assembled in compact fashion. Simulation results for transverse magnetic fields provided by this construction with 8.5 mm gap we show in Fig. 2. It is seen that on-axis peak field $B_0 = 1.36$ T is achievable.

Schematic view of the helical hybrid undulator is given in Fig. 3. Simulations of the field were made with the respective component dimensions optimized for the same parameters $\lambda_w = 30$ mm and $g = 8.5$ mm. The magnetic field profiles produced by this construction are presented in Fig. 4. On axis peak fields $B_0 = 0.96$ T is obtainable for both field components.

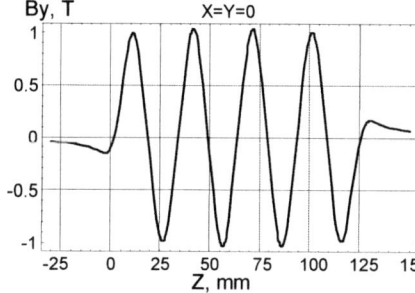 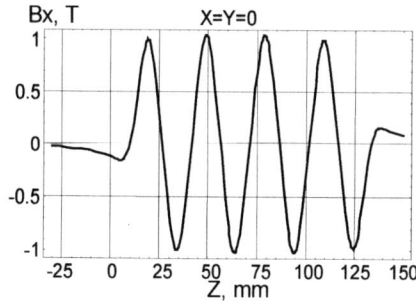

FIGURE 4. Transverse magnetic fields in the KIAE-3H undulator model. Undulator period $\lambda_w = 30$ mm, gap $g = 8.5$ mm, on-axis peak field $B_0 = 0.96$ T.

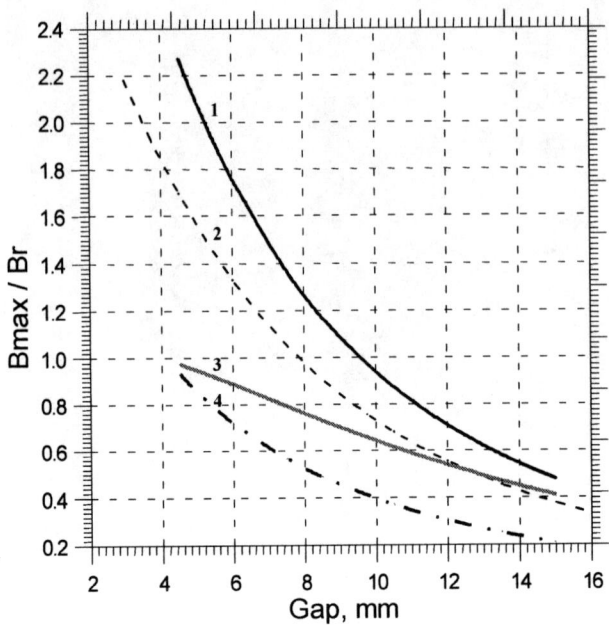

FIGURE 5. Peak magnetic fields provided by hybrid undulators with periods $\lambda_w = 30$ mm.
   1 - Novel planar KIAE-3P undulator scheme.
   2 - Routine hybrid planar scheme of Halbach [3].
   3 - Novel helical KIAE-3H undulator scheme.
   4 - Helical undulator scheme of Japan group [4].

Peak magnetic fields available with the above undulator schemes are shown in Fig. 5 as functions of gaps. Results for routine planar Halbach scheme [3] and results for one of the best hybrid helical undulator construction [4] are given also for comparison. It is seen that using permanent magnet material of the same permeability B_r one can obtain 35% higher fields with the described planar undulator scheme KIAE-3P and 80% higher field in the described helical undulator scheme KIAE-3H respectively than the maximum field strength achievable with known before hybrid undulator schemes.

CONCLUSION

With the considered above novel schemes for undulators required by LCLS project [1] high strength fields can be obtained at larger than was admitted before gaps and namely at 8.5 mm gap. Both options of undulators, planar and helical ones, can provide the specified field with this gap. So improvements [2] of the X ray FEL option [1] become possible.

REFERENCES

1. *LCLS Design Study Report,* Report SLAC-R-521 (1998).
2. Pellegrini C., Reiche S., Rosenzweig J., Schroeder C., Varfolomeev A., Tolmachev S., Nuhn H. "Optimization of an X ray SASE-FEL" *22-nd International Free Electron Laser Conference and 7-th FEL Users Workshop, 14 -18 August 2000, Durham, NC USA.*
3. Brown D., Halbach K., Harris J., Winick H., *Nucl. Instr. and Meth.* ***208,*** *65-67 (1983).*
4. Tsunawaki Y., Okuda Y., Ohigashi N., Kusaba M., Imasaki K., Fujita M., Mima K., Nakai S. "Development of a Hybrid Helical Microwiggler with Four Poles per Period" Proceedings *Free Electron Lasers 1999,* edited by J.Feldhausand H.Weise, 2000 Elsevier Science B.V., pp. II-91-II-92.

Numerical Study of SASE Mode FEL Evolution Starting from Noise with Subsequently Induced Slowing Down of Optical Pulse.

A.A. Varfolomeev and T.V. Yarovoi

CRL, Russian Research Center "Kurchatov Institute"
123182 Moscow, Russia

Abstract. A SASE mode FEL evolution starting from noise was simulated using an 1D code. Temporal and space structure of the optical pulse as well as the electron bunch were numerically studied. New possible effects from slowing down of spikes and from phase jumps of the optical pulse were investigated. It is shown that slowing down of the spikes before saturation takes place can enhance the output field intensity by 10%. Along with this a phase jump near the saturation depth can give intensity increasing by 50%.

INTRODUCTION

Spiking behavior of high-gain FELs is a well known phenomena [1-3]. Spikes can evolve in the so called steady state regime for long bunches in the slippage region [1] mainly at undulator depth distances exceeding the saturation length. Different results are obtained for SASE FELs starting from chaotic noise [2,3]. In this case the spikes appear at smaller depths before saturation. It is a more significant result from practical point of view since for all projected FEL SASE installations undulators not longer than the first saturation depth are planned. For short wave-length FELs including X-ray FELs, no primary bunching or feeding is available. So starting from noise is typical for the contemporary SASE FEL projects for generating short-wave length radiation.

In the FELs starting from noise, primary electron beam (e.b.) density nonuniformity leads to a nonuniform radiation wave packet with spikes. The radiation pulse with spikes induces nonuniform microbunching of the electron beam with its own spikes of microbunching. The electromagnetic (e.m.) spikes and maxima of the microbunched e.b. structure move with different velocities. Because of slippage, relative positions of the e.m. spikes and of the respective microbunched density spikes are not conserved.

The idea of the present work is to increase the time of correlated positions by slowing down in average the radiation spikes. The better overlapping of the field with

CP581, *Physics of, and Science with, the X-Ray Free-Electron Laser*, edited by S. Chattopadhyay et al.
© 2001 American Institute of Physics 0-7354-0022-9/01/$18.00

the microbunch structure should provide more efficient beam interaction and enhance the radiation gain*. The above mentioned slowing down of the e.m. wave packets can be provided with using special insertion devices suggested and described in our previous work [5].

Another effect which will be analyzed is a local phase jump influence on the SASE FEL dynamics [6]. Relative phase shifts of electrons can be induced by a short drift space in the intersection gap [7]. Up to date investigation [6] was limited to the steady state regime. It will be shown in this paper that both the above effects can be used for the enhancement of SASE FEL output power.

EQUATIONS AND A MODEL SHEME USED IN THE SIMULATIONS

For simplicity a single pass high gain FEL is considered in 1D approach and in a slow-varying envelope approximation (SVEA). The undulator sections are supposed to be of constant parameters. The respective slippage compensation and the phase jumps are introduced directly at some depths supposing that in practice it can be done by the special insertion devices positioned between undulator sections.

A simple specially written one-dimensional code was used for the study of superradiant effects. A peculiarity of this code is the possibility to make a slippage compensation and phase shifting for the optical pulse. Similar to other works (see for example [2]) the electron beam was "sampled" at each radiation wave-length λ_s. N_p model particles or simulation electrons were used. For tracing each model particle j along an undulator period length the following coupled evolution equations were used:

$$\frac{\partial \theta_j}{\partial z} = p_j$$

$$\frac{\partial p_j}{\partial z} = -\left(A \cdot \exp(i\theta_j) + c.c.\right) \qquad (1)$$

$$\frac{\partial A}{\partial z} = \langle \exp(-i\theta) \rangle$$

where notations similar to that of [1] were used and namely: $z = \frac{4\pi}{\lambda_w} \rho z$ is position of simulation electron expressed in the gain length, θ_j – is the phase of this simulation electron, $p_j = \frac{\gamma_j - \langle \gamma \rangle_0}{\rho \langle \gamma \rangle_0}$ is the energy deviation from the average electron energy; $A = E/(4\pi \rho n_e \langle \gamma \rangle_0 mc^2)^{1/2}$ is the dimensionless field strength, n_e is the electron

---

* Effect of moving spikes along the electron bunch improving FEL characteristics was considered also by Pellegrini [4]. Instead of this radiation spikes superimposing [4], we consider superimposing of the radiation spikes with electron microbunches.

beam density, $\langle \gamma \rangle_0 mc^2$ is average electron initial energy, ρ is the fundamental FEL parameter. The symbol $\langle ... \rangle = \frac{1}{N_p} \sum_{j=1}^{N_p} ...$ means averaging on N_p sample particles, $\langle \exp(i\theta) \rangle \equiv b$ is the longitudinal e.b. bunching.

For taking the slippage into account, the radiation wave packets after every one step of depth λ_w were moved ahead a λ_s distance with respect to the electron bunch i.e. on the next λ_s section of the e-beam. The simulation electrons were first spaced uniformly along the λ_s wave-length (with uniform distribution of phases θ_j within λ_s section). Then their phases were redistributed by adding small random additional phases $\delta\theta_j$ distributed with a Gaussian function of width $\Delta=(N_p/N_\lambda)^{1/2}$ where N_λ is the actual number of real electrons on λ_s length defined by the electron beam current. In this way a correct modulation of the random noise can be simulated for the average bunching parameter $\langle |b_0|^2 \rangle \cong \frac{1}{N_\lambda}$.

NUMERICAL SIMULATION RESULTS IN LONG ELECTRON PULSE LIMIT

The above code and the simulation scheme were first tested by simulations made for the parameters used in the calculations [1]. The same result were obtained for the intensity $|A|^2(z_1)$ (for depths z respectively of 10, 20, and 30 with given parameters of [1]: K=1/30, S=1, A_0=0.01, b_0=0). For another test the simulations were made for an uniform primary bunching (not random) along the bunch length $l_b=15l_c$. Here $l_c=4\pi\lambda_s/\rho$ is the cooperation length. At the saturation depth z_{sat} a local phase jump equal to π was introduced. Obtained results shown in Fig. 1 are identical to the

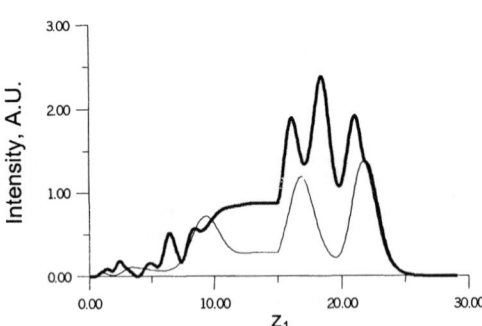

FIGURE 1. Field intensity $|A|^2$ as a function of the distance from the bunch tail expressed in l_c units $z_1=z_b/l_c$.

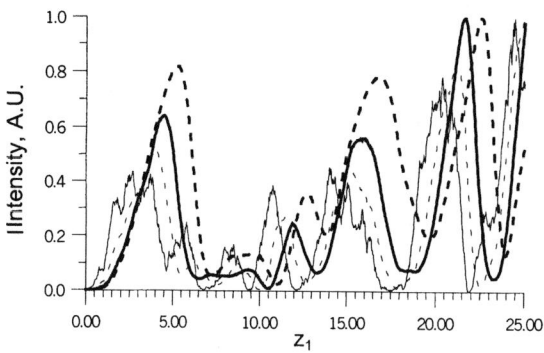

FIGURE 2. Normalized field intensities as a function of the distance along the bunch z_1 at different depths for normal FEL evolution. Thin curve - z =4, thin dash curve - z =6, bold curve - z =8, curve dash line - z =10.

respective results of Ref. [6] where the phase jump effect was investigated for the steady state regime for a long bunch $l_b >> l_c$.

The main simulations of the SASE mode FEL were made for a bunch length $l_b=25l_c$ and an average primary bunching $|b_0|=0.01$. Fig.2 shows spatial and temporal evolution of the radiation spike structure. Here z_1 is a distance from the bunch tail expressed in the cooperation length l_c. Radiation intensities relating to different depths z are normalized to have equal maximal spike peaks. Number of spikes is approximately $l_b/2\pi l_c$ as was predicted by the SASE FEL theory [1-3]. It is evident that spike shapes become more smooth along the depth z increasing, and that they move ahead with an average slippage ~0.36 l_c/gain length.

To improve overlapping of the radiation and bunching spikes, the radiation wave packets were displaced back on 0.75 l_c every time when the wave packet passed a gain length. This operation was repeated up to the depth z =9 inclusively. As a result

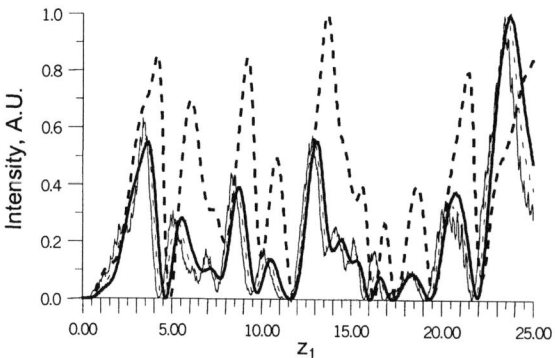

FIGURE 3. Normalized field intensities as a function of z_1 at different depths for regime with wave packet displacements. Thin curve - z =4, thin dash curve - z =6, bold curve - z =8, bold dash curve - z =10.

the spike structure given by Fig. 3 was obtained. It is seen that the radiation spikes have now very small average slippage with respect to the e.b. bunch at $z \leq 9$ and go ahead again at $z = 10$ where the displacement was not made since the saturation is nearly reached and the slippage compensation does not give a noticeable gain enhancement (see below).

The phase shift $0.75\ l_c$ was found to be optimal in these simulations. As one can see it is not equal to the slippage found for the normal FEL regime. This difference can be explained by the larger influence of a spike lethargy effect in the normal operating high gain FEL than in the case of the compensated slippage regime.

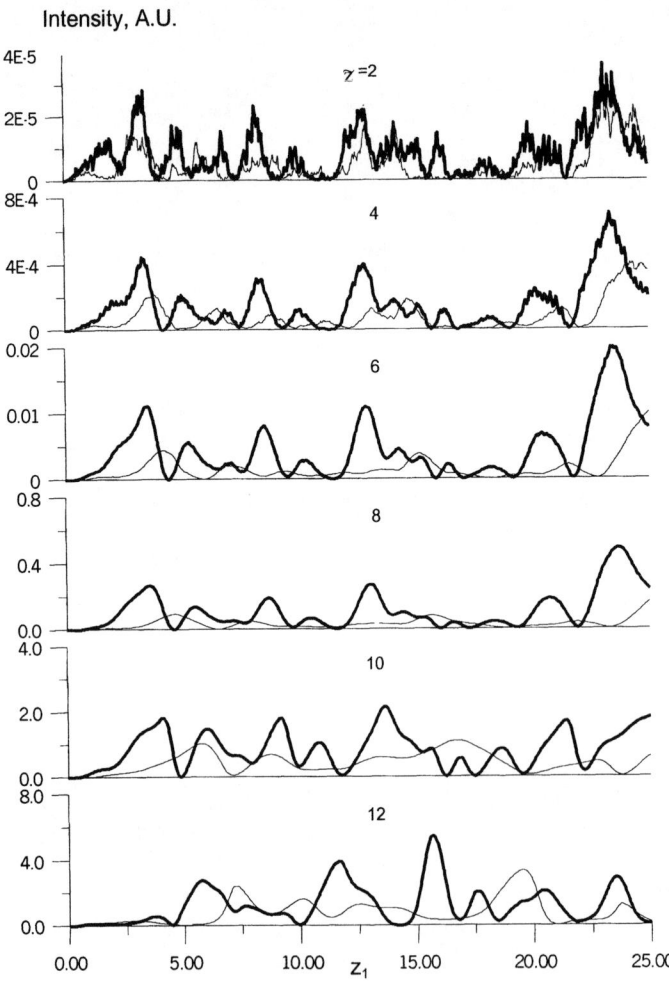

FIGURE 4. Field intensity $|A|^2$ distributions at different depths z expressed in absolute units. Thin curve – no slippage compensation, bold curve – optical pulse displacement on $0.75l_c$ at every gain length up to $z = 9$ inclusively.

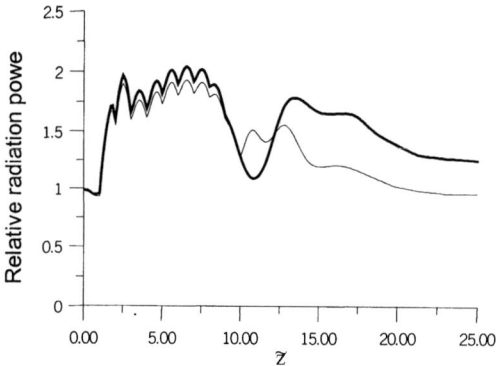

FIGURE 5. Relative output radiation power normalized by the value of the output radiation power from a normal FEL evolution. Bold curve – slippage compensation is made, Thin curve – both the slippage compensation and the phase shift $\pi/2$ are made.

Radiation intensities in absolute units obtained in single runs with the same primary noise distribution are given in Fig. 4. The peaks are growing in time and changing their shapes. It is clearly seen that slippage compensation provides higher radiation intensities. Fig. 5 shows average results for the ratio of the total field intensities. Results have been obtained from 20 simulation runs. It is seen that the induced slippage compensation doubled the radiation intensity in the linear regime of the SASE mode at $z < z_{sat}$. The averaged output radiation intensities in absolute units as functions of the depth are presented in Fig. 6. These data have also been obtained

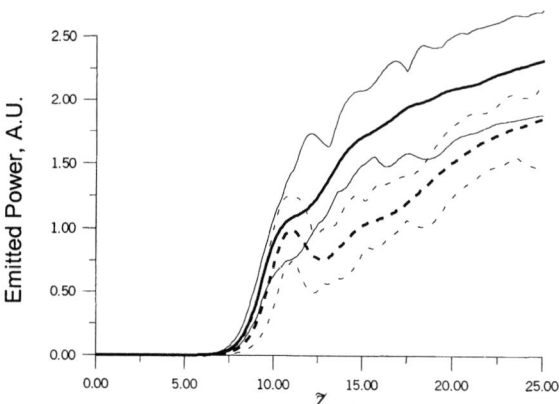

FIGURE 6. Output radiation power as a function of the dimensionless distance along the undulator. Bold curve – $0.75 l_c$ displacement every one gain length up to $z = 9$, bold dash curve – no induced displacement; thin curve and thin dash curve – the respective fluctuation limits.

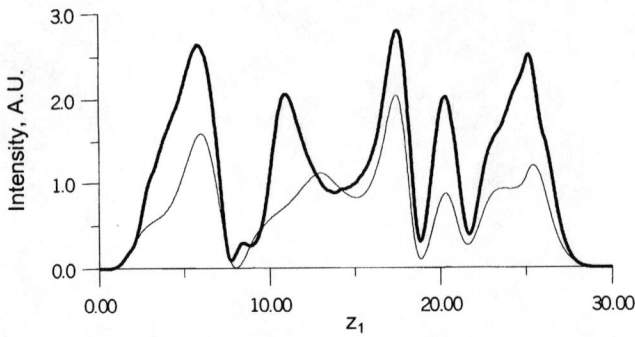

FIGURE 7. Field intensity $|A|^2$ at the depth $\tilde{z}=11$. Thin curve – basic regime of normal FEL operation (usual undulator), bold curve – phase jump $\pi/2$ induced at $\tilde{z}=10$, dash curve – intensity distribution at the depth $\tilde{z}=10$ (for comparison).

from 20 simulation runs. At the usual saturation depth the intensity increased by 10% due to the above slippage compensating displacements but it is going on to grow at larger depths without any decreasing. enhancement of 25% is possible at the depth $\tilde{z}=12.5$.

As was mentioned earlier another effect we have investigated is the local phase jump induced at the saturation depth near $\tilde{z}=10$. Fig. 7 demonstrates how this phase shifts can change the spike structure. All spikes have been grown along the whole bunch length. The reason of this effect can be understood from Fig. 8 where phase space plots are given for a point $z_1=15$ at different depths $\tilde{z}$, a little less and greater than the saturation depth respectively. It is seen that the phase shift at $\tilde{z}=10$ induces additional detrapping of the electrons from the pondermototive bucket bottom (when the electrons have lost energy). Thus the output radiation intensity can be enhanced not

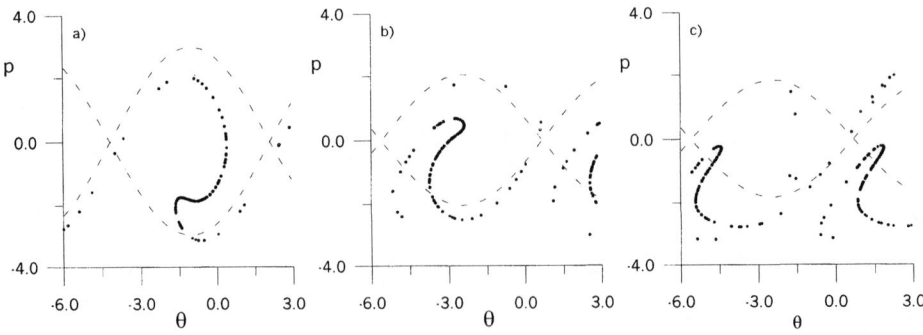

FIGURE 8. Phase space plots of electrons at the point $z_1=15$. a) at the depth $\tilde{z}=10$, with no phase shift, b) at the depth $\tilde{z}=12$, no phase shift induced, c) at the depth $\tilde{z}=12$, phase jump $\pi/2$ at $\tilde{z}=10$ induced.

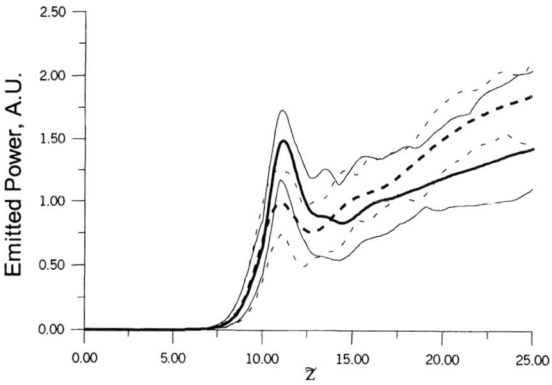

FIGURE 9. Output radiation power as a function of the undulator depth z. Bold curve – averaged power with $\pi/2$ phase shift induced at the depth $z=10$, thin curves – fluctuation limits for bold curve results, bold dash – averaged results with no induced phase shift, thin dash curves – fluctuation limits for bold dash results.

only in the steady state regime [6] but also with limited bunch lengths and random noise primary bunching. As Fig. 9 shows this local phase jump can enhance the output intensity by 50%.

Simulations were made also for the case when both the slippage compensation and the local phase jump were induced. The total gain in the output intensity is not changed much in comparison with that in the phase jump case. So both these effects do not interfere much in the average radiation power production (See Fig. 10).

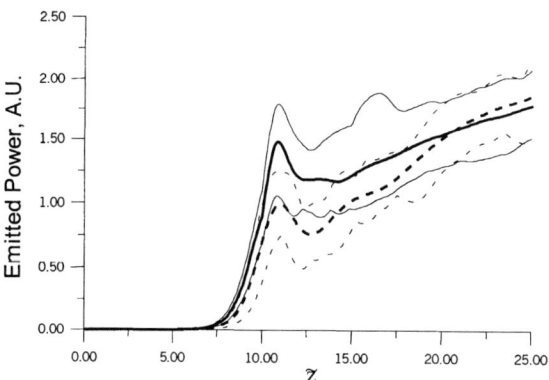

FIGURE 10. Output radiation power as a function of the depth z. Bold curve – averaged power with $0.75l_c$ displacement every one gain length up to $z=9$ followed by phase shift $\pi/2$ induced at the depth $z=10$, thin curves – fluctuation limits for the bold curve results, bold dash – averaged results with no induced phase shift, thin dash – fluctuation limits for the bold dash results.

CONCLUSION

The numerical simulation of the SASE FEL process starting from noise for a long bunch ($l_b=25l_c$) did show that both the average slippage compensation as well as the local phase displacement at the saturation depth enhance the output radiation power. The local phase jump effect is more efficient and works similarly to that found for the steady state regime. The slippage compensation effect can be used for additional enhancement as well. The total intensity enhancement of up to 50% is demonstrated in these simulations.

REFERENCES

1. Bonifacio, R., De Salvo Souza, L., Pierini, P., and Piovella, P., *Nucl. Instr. and Meth.* **A296** (1990) 358.
2. Bonifacio, R., De Salvo, L., Pierini, P., Piovella, N., and Pellegrini, C., *Nucl. Instr. and Meth.* **A341** (1994) 181, *Phys. Rev. Lett.* A3 (1994) 70.
3. Bonifacio, R., Pierini, P., Pellegrini, C., Rosenzweig, J., and Travish, G., *Nucl. Instr. and Meth.* **A341** (1994) 285.
4. Pellegrini, C., *Nucl. Instr. and Meth.* **A445** (2000) 124.
5. Varfolomeev, A.A., and Yarovoi, T.V., *Nucl. Instr. and Meth.* **A393** (1997) 398.
6. Varfolomeev, A.A., Yarovoi, T.V., and Bousine, P.V., *Nucl. Instr. and Meth.* **A407** (1998) 296.
7. Varfolomeev, A.A., and Yarovoi, T.V., *Nucl. Instr. and Meth.* **A375** (1996) 352.

Velocity Bunching in Photo-Injectors

L. Serafini and M. Ferrario*

University of Milan and INFN-Milan
* INFN-Frascati

Abstract. We describe here a new method to increase the peak current of high brightness electron beams as those required to drive X-ray SASE FEL's, that is based on a rectilinear compressor scheme utilizing the bunching properties of slow waves. It is shown that whenever a beam, slower than the synchronous velocity, is injected into a RF wave at the zero acceleration phase and slips back in phase up to the peak acceleration phase, it can be compressed as far as the extraction happens at the synchronous velocity. In fact, the bunch undergoes a quarter of synchrotron oscillation that induces a net compression (i.e. a bunch length reduction) up to a factor of 20 when proper care is taken to preserve the longitudinal emittance. A few examples are presented to demonstrate the potentialities of this method, by which multi-kA beams at very low emittance can be generated at moderate energies (about 100 MeV).

INTRODUCTION

The need to produce high brightness electron beams delivered in short (sub picosecond) bunches has been driven recently by the demands of X-ray SASE FEL's, which typically require multi-GeV beams with multi-kA peak currents and bunch lengths in the 100-300 fs range, associated to normalized transverse emittances lower than a few mm·mrad.

The foreseen strategy to attain those beams is based in present designs on the use of RF Linacs in conjunction with RF laser driven Photoinjectors and magnetic compressors. The formers are needed as sources of low emittance high charge beams (usually 2 mm·mrad at 1 nC bunch charge) with moderate currents (< 100 A @ 10 MeV), the latter are used to enhance the peak current of such beams up to the design value of 2-3 kA by reduction of the bunch length achieved at relativistic energies (> 300 MeV, up to 5 GeV in multi-staged compressors).

Since the impact of magnetic compressors on the beam dynamics is quite relevant, with tendency to jeopardize the performances of the whole system in terms of the final beam brightness achieved – this makes these type of compressors one of the critical components of these Linacs – we tried to explore an alternative option of compression (i.e. of current enhancement) based on slow wave RF fields.

The performances of this option are presented in this paper on the basis of a preliminary analysis, whose main conclusion is that this method of compression has very promising potentialities and is a good candidate to substitute magnetic compressors, alleviating in this way the problems related to Coherent Synchrotron Radiation emission occurring in the chicanes of magnetic compressors and its critical impact on the beam quality (brightness) achievable.

CP581, *Physics of, and Science with, the X-Ray Free-Electron Laser*, edited by S. Chattopadhyay et al.

The great advantage of the RF rectilinear compressor scheme presented here is obviously the absence of curved path trajectories in the compressor, in addition to the fact that compression is applied at moderate energies (from 10 to 100 MeV), leaving the Linac (up to the final energy) free from any further beam manipulation needed to increase the current.

The rectilinear RF compressor works indeed as a standard accelerating structure which, while accelerates the beam, it also reduces its bunch length.

We have to acknowledge here that the seed of the idea leading to the RF compressor was set in the development of a previous work on a plasma buncher scheme[1] which turned out to be of great relevance for the studies on second generation plasma accelerators[2].

In that work indeed it was shown that a plasma accelerator working at some resonant gamma γ_r can accelerate and simultaneously compress long bunches injected at a proper phase, i.e. the phase of no net acceleration. In that specific example, a plasma wave driven at an amplitude of 3 GV/m with plasma wavelength of 300 μm accelerates an electron bunch injected at 2.5 MeV up to an energy of 200 MeV, by simultaneously reducing the bunch length from 15 μm at injection down to 3 μm . In order to achieve this effect is necessary to inject at 0° phase (no acceleration) so to let the beam slip back in phase and extract the beam when its energy is close to the resonant gamma of the wave (in this case $\gamma = 400$). In case of injection at 90° (max acceleration) the bunch would be accelerated up to a higher final energy (290 MeV) but its length would be left unchanged.

The longitudinal component E_z of the RF field in a traveling wave structure or a Langmuir wave in a cold plasma can be well approximated by a simple expression neglecting higher order spatial harmonics

$$E_z = -E_0 \sin(\omega t - kz + \psi_0) \qquad (1)$$

where the wave number k is taken to be $k = \dfrac{\omega}{\beta_r c}$ (implying a wave phase velocity given by $v_r = \beta_r c$, $\gamma_r = 1/\sqrt{1 - \beta_r^2}$) and the wave amplitude E_0 can be cast in terms of a dimensionless parameter α , defined as $\alpha \equiv \dfrac{eE_0}{mc^2 k}$, which represents the normalized amplitude of the vector potential associated to the wave electric field (we are considering here waves with phase velocity quite close to c, or, in other words, resonant gamma's quite large, $\gamma_r \gg 1$).

The interaction between an electron and a wave of this kind can be described by an Hamiltonian

$$H = \gamma - \beta_r \sqrt{\gamma^2 - 1} - \alpha \cos \xi \qquad (2)$$

where γ is the normalized energy of the electron, $\gamma = 1 + \dfrac{T}{mc^2}$ and $\xi = kz - \omega t - \psi_0$ the phase of the wave as seen by the electron.

Let us consider now a wave whose phase velocity is slightly smaller than c, so that $k = k_0 + \Delta k = \dfrac{\omega}{c} + \Delta k$, and the detuning parameter Δk is small, i.e. $\Delta k \ll k_0$; the resonant beta and gamma can be well approximated by

$$\beta_r \cong 1 - \frac{\Delta k}{k_0} = 1 - \frac{c\Delta k}{\omega} \quad \text{and} \quad \gamma_r \cong \sqrt{\frac{k_0}{2\Delta k}} = \sqrt{\frac{\omega}{2c\Delta k}} \tag{3}$$

In previous example related to the plasma wave accelerator we have $\gamma_r = 400$, so the detuning parameter is $\Delta k = 3.1 \cdot 10^{-6} k_0$.

The trapping and acceleration in these waves has been widely discussed in the literature (see for example in ref.3), and it can be well understood looking at the phase contour plots in the $[\gamma, \xi]$ phase space. One of the most relevant feature is that whenever the resonant gamma is finite, i.e. the wave is slower than c, we have a separatrix in this space, splitting trapped orbits from untrapped ones.

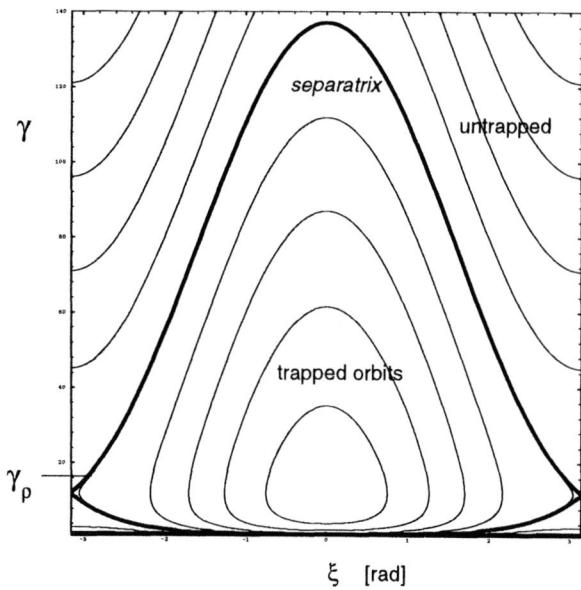

FIGURE 1. Phase space plots for a slow wave with resonant gamma $\gamma_r = 12$ and $\alpha = 0.2$. The separatrix is shown by the bolded line.

This situation is illustrated in Fig.1 for the case of a wave of amplitude $\alpha = 0.2$ and resonant gamma $\gamma_r = 12$. This corresponds to a wave phase velocity of $\beta_r = 0.9965$ or an equivalent detuning parameter of $\Delta k = 3.5 \cdot 10^{-3} k_0$. The top of the separatrix (i.e.

the value of γ at $\xi = 0$ for $\gamma_0 = \gamma_r$ and $\psi_0 = \pi$) is approximately given by $\gamma_{max} = 4\alpha\gamma_r^2$, which is the maximum energy of trapped particles.

Since the Hamiltonian written in eq.2 is an invariant of the motion, we can evaluate it at injection, getting:

$$\gamma - \beta_r \sqrt{\gamma^2 - 1} - \alpha\cos\xi = \gamma_0 - \beta_r \sqrt{\gamma_0^2 - 1} - \alpha\cos\psi_0 \qquad (4)$$

where γ_0 and ψ_0 are the initial energy and phase of the particle injected into the wave. The maximum accelerating gradient occurs at $\xi = -\pi/2$: if particles are injected at this phase inside the separatrix region with an initial energy γ_0 smaller than the resonant one γ_r they will slip back in phase but they will be accelerated up to $\gamma = \gamma_r$, where their velocity will match that of the wave, thereafter they will start to slip ahead of the wave during the acceleration up to the highest γ value of their phase contour line (well above γ_r) corresponding to $\xi = 0$. At this point they will start being decelerated unless extracted from the wave.

There is one relevant property of this kind of energy oscillation performed into the bucket by trapped electrons, which turns out to be very helpful in reducing the phase distance of particles injected close to each other into the wave at small energy.

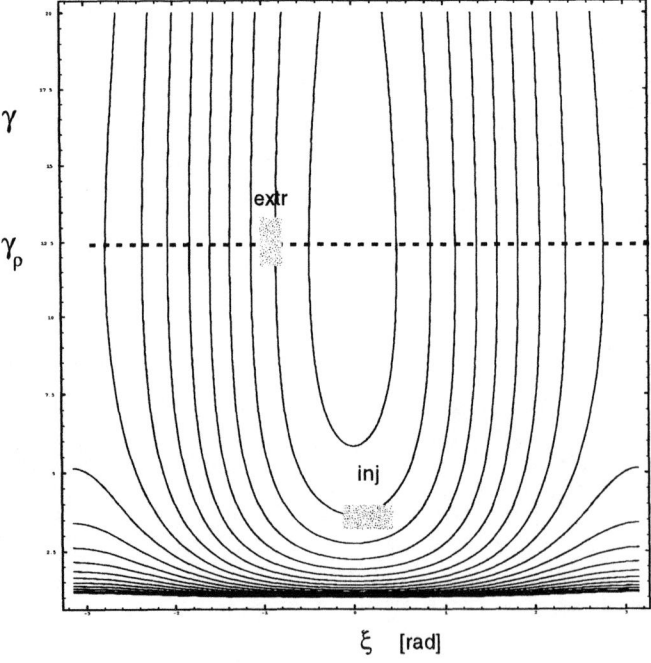

FIGURE 2. Enlarged part of the diagram plotted in Fig.1, corresponding to $\gamma \leq 20$.

Because of the shape of the phase contour lines, the region around $\xi = 0$ corresponds to high phase spreads and low energy spreads while the opposite is true for the region around $\xi = -\pi/2$ where the phase lines are parallel to the γ-axis.

This is quite well illustrated in Fig.2 where an enlarged part of the diagram plotted in Fig.1 is shown, and exactly that corresponding to $\gamma \leq 20$. In this figure a particle distribution at injection is depicted in a small phase space area located around $\xi = 0$ and $\gamma_0 = 3$, i.e. well below γ_r (i.e. particles initially slower than the wave) and close to the bottom of the cup-shaped bucket. This bunch of particles is accelerated by the wave while it slips back in phase: once at the resonant energy γ_r it will reach a phase of about $\xi = -60°$. As depicted in the diagram, due to the nature of the phase lines, the bunch will have a phase spread (i.e. a bunch length since it is quite relativistic at this point) smaller than the initial one, which means that this kind of acceleration acts also as a phase compression.

Extracting the bunch when it reaches γ_r is therefore the way to achieve a bunch length compression at the same time as acceleration: a further acceleration beyond γ_r would clearly tend to de-compress the bunch, as clearly shown by the shape of the phase lines above γ_r (inside the separatrix).

We can evaluate the amount of phase compression achieved by assuming injection at $\xi = 0$ and extraction at $\gamma = \gamma_r$, so that eq.4 becomes

$$\frac{1}{\gamma_r} - \alpha \cos \xi_{ex} = \gamma_0 - \beta_r \sqrt{\gamma_0^2 - 1} - \alpha \qquad (5)$$

which basically determines the extraction phase ξ_{ex} as a function of the injection conditions (γ_0, ψ_0) and the wave parameters (γ_r, α).

We also assume that the injection phase ψ_0 and energy γ_0 will be distributed around some average values $\bar{\gamma}_0$ and $\bar{\psi}_0$ with rms spreads defined as $\gamma_0 = \bar{\gamma}_0 \pm \delta\gamma_0$ and $\psi_0 = \bar{\psi}_0 \pm \delta\psi_0$. These initial spreads in energy and phase at injection will affect the final phase spread $\delta\xi_{ex}$ at extraction as given by $\alpha\cos(\bar{\xi}_{ex} \pm \delta\xi_{ex}) = 1/\gamma_r - (\bar{\gamma}_0 \pm \delta\gamma_0) + \beta_r \sqrt{(\bar{\gamma}_0 \pm \delta\gamma_0)^2 - 1} + \alpha\cos(\pm\delta\psi_0)$, which can be cast, through a second order expansion, in the form

$$\delta\xi_{ex} = \frac{1}{2|\sin\bar{\xi}_{ex}|} \sqrt{\delta\psi_0^4 + \left(\frac{1}{\alpha\bar{\gamma}_0} \frac{\delta\gamma_0}{\bar{\gamma}_0}\right)^2 \left(\frac{\bar{\gamma}_0^2}{\gamma_r^2} - 1\right)^2} \qquad (6)$$

where the extraction phase is specified by $\cos\bar{\xi}_{ex} = 1 + \frac{1}{\alpha\gamma_r} - \frac{\bar{\gamma}_0}{\alpha}(1 - \beta_r\bar{\beta}_0)$.

The relevant parameter is the compression factor C, sort of a figure of merit for this process, defined as the ratio between the initial phase spread and the final one at extraction, i.e. $C \equiv \frac{\delta\psi_0}{\delta\xi_{ex}}$, which turns out to be given by

$$C = \frac{2\delta\psi_0 |\sin \overline{\xi}_{ex}|}{\sqrt{\delta\psi_0^4 + \left(\frac{1}{\alpha\overline{\gamma}_0} \frac{\delta\gamma_0}{\overline{\gamma}_0}\right)^2}} \qquad (7)$$

where we assumed $\frac{\overline{\gamma}_0^2}{\gamma_r^2} \ll 1$.

Taking the example shown in Fig.2, i.e. a bunch injected at $\overline{\gamma}_0 = 3$ with $\frac{\delta\gamma_0}{\overline{\gamma}_0} = 5 \cdot 10^{-3}$ and $\delta\psi_0 = 5 \cdot 10^{-2}$ into a wave specified by $\alpha = 0.2$ and $\gamma_r = 12$, we derive from eq.7 a compression factor $C = 9.8$, at an extraction phase $\overline{\xi}_{ex} = -60^o$, which indicates a quite interesting capability to increase the peak current in the bunch by a relevant factor (in excess of 10).

Taking as a second example the one considered in the work on plasma bunchers, i.e. a bunch injected at $\overline{\gamma}_0 = 6$ with $\frac{\delta\gamma_0}{\overline{\gamma}_0} = 5 \cdot 10^{-3}$ and $\delta\psi_0 = 0.31$ (a bunch 15 μm long into a plasma wave with 300 μm wavelength) into a wave specified by $\alpha = 0.28$ (3GV/m amplitude) and $\gamma_r = 400$, we derive from eq.7 a compression factor $C = 4.5$, at an extraction phase $\overline{\xi}_{ex} = -45^o$, which is very close to the factor 5 in bunch length reduction observed in the numerical simulations, that foresee a final bunch length of 3 μm at the exit of the plasma wave[1] .

RF RECTILINEAR COMPRESSORS

On the basis of the bunch compression mechanism described above we can envisage a very useful application of this concept in the frame of high brightness photoinjectors for X-ray FEL drivers.

The aim of the idea is to try compressing the beam right after the exit from the RF gun, when the energy is about 5-7 MeV, by injecting the bunch into an accelerating structure which supports a slow wave like the one specified in eq.1, so that the beam would leave this structure at the final maximum value of peak current achievable at an energy close to (or somewhat higher) than the resonant gamma. If this exit energy is high enough we can envisage that the residual longitudinal space charge field will not debunch the beam further on, in particular if the beam is promptly injected into a regular high gradient accelerating structure in order to relativistically freeze down the residual - generally weak – longitudinal space charge forces.

We will consider here the lay-out of the injector for the LCLS[4], which is one of the most promising designs for high brightness beams recently proposed[5], based on the use of S-band traveling wave structures boosting a 6.5 MeV beam produced by the RF gun. The paradigm for this beam is: 1 nC bunch charge, 10 ps laser pulse length on the photocathode, 140 MV/m peak field on the photocathode, resulting in a 100 A peak current at the gun exit with 0.8 % energy spread and 0.5 mm mrad rms normalized transverse emittance (as predicted by simulations). A solenoid lens located right at the gun exit focuses down the beam in order to match its envelope onto the invariant envelope[7] at the injection in the first TW structure, which is located 1.5 m far from the cathode[5]. We just take this lay-out and substitute the first TW structure with one that is properly designed to support a slow wave with wave number 0.07 % higher than the nominal one (corresponding to a cell length 0.07 % shorter than the standard $\lambda/3$ nominal value of a $2\pi/3$ mode – this perturbative change in wave number is equivalent to a 2 MHz detuning of the structure, i.e. to having a 2854 MHz field propagating through a standard S-band SLAC structure). This structure will have 86 cells over a length of about 3 m, operating at 2856 MHz.

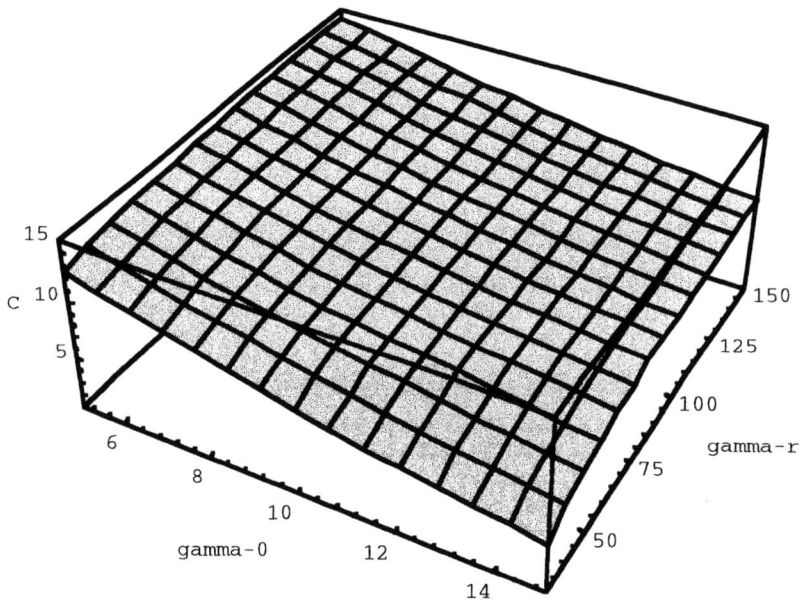

FIGURE 3. Compression factor C as a function of the initial energy γ_0 and the resonant gamma γ_r for a wave of amplitude $\alpha = 0.75$ and an injected bunch characterized by $\delta\gamma_0/\bar{\gamma}_0 = 8\cdot10^{-3}$ and $\delta\psi_0 = 5\cdot10^{-2}$ (LCLS parameters at the gun exit).

As shown in Fig.3, the compression factor C for a wave of amplitude $\alpha = 0.75$ (corresponding to an accelerating gradient of 23 MV/m at S-band) is a decreasing function of γ_0 for a given γ_r and increases slightly with γ_r for a given γ_0.

Since in the LCLS injector case the initial energy is $\gamma_0 = 14$ and the available TW structures are 3 m long, we picked up the following optimum parameters for the slow wave structure: $\gamma_r = 27$, corresponding actually to the previously mentioned 0.07 %

increase in wave number and $\alpha = 0.75$. For this set of parameters the predicted compression factor is $C = 6.5$.

By optimizing the injection phase, we ran Homdyn [6] from the cathode surface all the way through the gun, drift, slow wave structure, up to the exit of a second 3 m long TW structure (regular one, no detuning, driven at 27 MV/m accelerating gradient), located right after the first detuned one. At the optimum phase (slightly earlier than 0°) we found a very nice final peak current of 870 A at an energy of 120 MeV, as shown in Fig.4.

The effect of the RF rectilinear compressor (the detuned structure), extending from z=1.5 m up to z=4.5 m, is clearly illustrated: the energy at its exit is 18.5 MeV ($\gamma = 37$, somewhat larger than γ_r) and the compression takes place almost entirely during the acceleration through it. The compression factor achieved is $C = 8.1$, a bit larger than the analytical prediction[10].

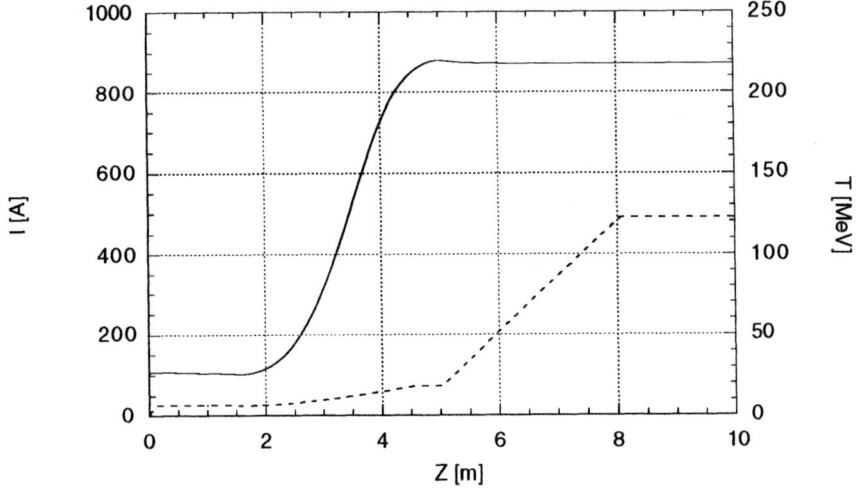

FIGURE 4. Beam current (solid line, left scale, in A) and energy evolution (dashed line, right scale, in MeV) through the LCLS injector with implementation of a RF rectilinear compressor.

The energy spread is plotted in Fig.5: after a large peak due to the injection at 0° phase, where the dominant effect is to induce energy spread instead of accelerating the bunch, it is naturally damped by acceleration, reaching about 1% at the exit.

The transverse dynamics is strongly altered by the compressor because the beam is all the way space charge dominated in the transverse plane: the plasma oscillation shown by the beam envelope in Fig.5 must be carefully taken under control by the solenoids wrapped around the two accelerating sections (by proper tuning of their field amplitude) so to achieve an emittance correction scheme[7] that extends through the two accelerating structures up to 120 MeV. The final value reached, 0.5 mm·mrad, is consistent with the design value reported for the LCLS injector study, showing that the optimal emittance correction can be achieved even in presence of the RF compressor.

The rms longitudinal emittance (not shown in the figures) goes from 25 keV·mm at the exit of the RF gun up to 155 keV·mm at the exit of the RF compressor: we believe that this is due to the non-linear behavior of the accelerating field versus the accelerating phase that introduces a curvature in the longitudinal phase space. The contribution of the longitudinal space charge forces is indeed negligible throughout the RF compressor: in fact, repeating the simulation shown in Fig. 4-5 by switching off the space charge at the injection into the compressor we have obtained results almost identical to the previous case (actually the final value of the peak current is slightly smaller – 860 A – for the zero space charge case, as for the longitudinal emittance).

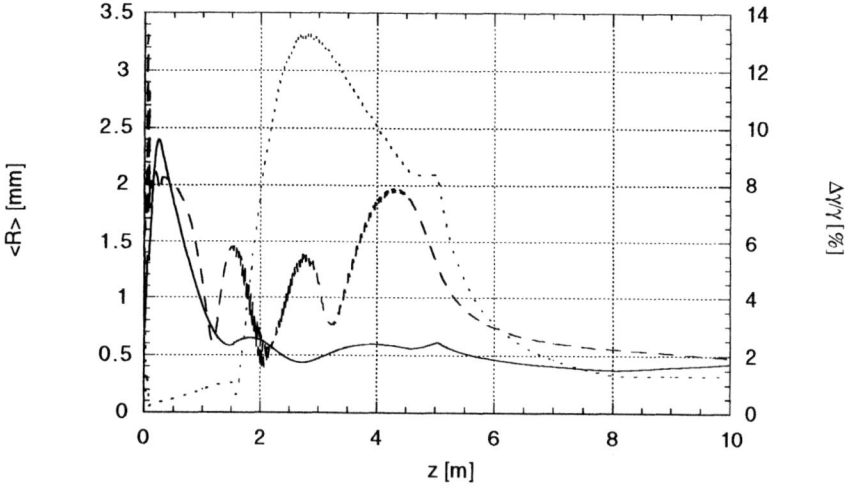

FIGURE 5. Energy spread in % (dotted line, right scale), beam envelope in mm (solid line, left scale) and rms normalized transverse emittance in mm·mrad (dashed line, left scale) evolution as for Fig.4.

Therefore, if the maximum peak current achievable is limited by RF non-linearities there should be two ways to overcome this problem: one is the decrease in the accelerating gradient the second is the use of higher harmonics to flatten-out the energy-phase correlation [8].

We will discuss the second option in the next section of this paper: let us now consider what is the behavior of the compression factor C as a function of α (i.e. the accelerating gradient). This is plotted in Fig.6, clearly displaying that there is actually an optimum value of α - about 0.25 - that maximizes the compression factor.

We repeated the simulations with the same injected beam and same slow wave structure as before, but changing now the gradient, i.e. applying 13 MV/m, corresponding to $\alpha = 0.44$, and increasing the structure length up to 6 m. The results are plotted in Fig.7: at an exit energy of 24 MeV the beam peak current is 1260 A with a longitudinal emittance of 80 keV mm. As anticipated, at lower gradients the longitudinal emittance is better due to lower non-linear RF effects, hence the maximum peak current achievable is considerably higher.

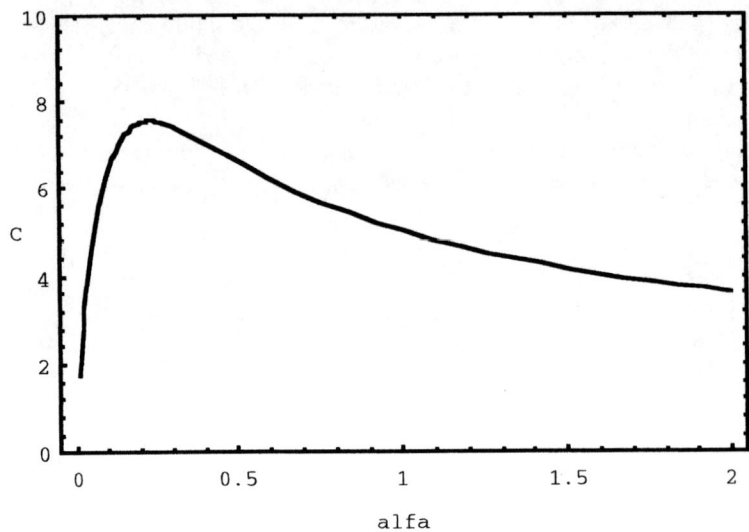

FIGURE 6. Compression factor C plotted as a function of α for the same beam conditions at injection and slow wave structure used in Fig.4-5.

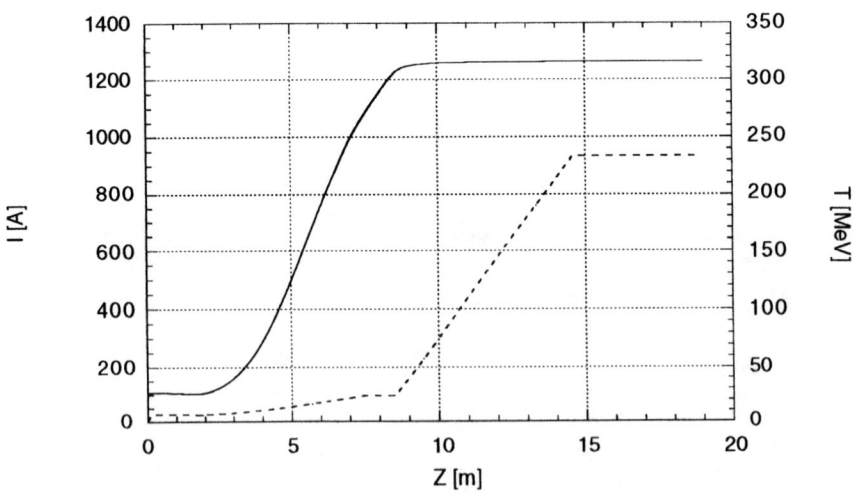

FIGURE 7. Beam current (solid line) and energy evolution (dashed line, right scale) through the LCLS injector with implementation of a 6 m long RF compressor driven at low accelerating gradient.

The impact of these RF non-linear effects is clearly shown by the behavior of the longitudinal phase space distributions plotted in Fig.8 at various locations through the compressor.

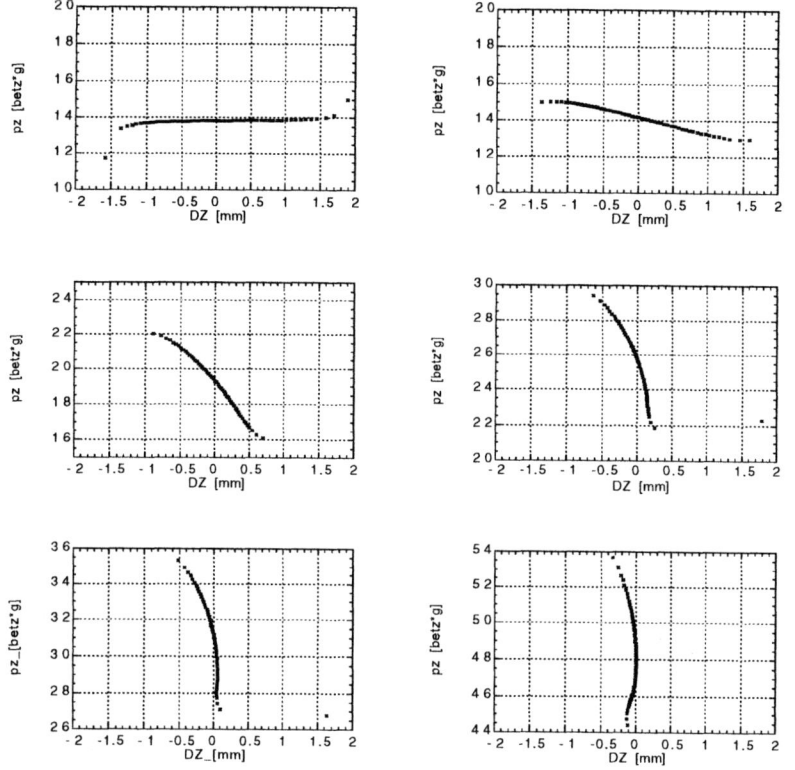

FIGURE 8. Beam phase space distributions in the $(\Delta z, p_z)$ longitudinal plane at different locations through the 6 m long RF compressor (these are snapshots taken in time, as the code Homdyn is time-dependent). First diagram is for the beam at injection, last diagram is for the beam right at the compressor exit.

The first diagram shows the energy phase distribution of the beam at injection into the compressor with a clear signature of the longitudinal space charge field action occurred from the photocathode surface up to the exit of the gun and further on through the drift to the compressor: some non-linearities are already present but mostly concentrated in the bunch tails, as expected. The second diagram is taken after the bunch has traveled a few wavelengths: since the injection is performed at $0°$, most of the RF field action goes into producing an almost linearly correlated energy spread with particles in the tail being at higher energy than those in the bunch head. After some more RF periods – third diagram – the bunch starts being accelerated because of the phase slippage to higher phases – the beam is indeed slower than the wave at this point. At the same time the beam starts being bunched because of the phase space rotation caused by the linearly correlated energy spread. These two actions – bunching and acceleration – proceed actually together in a synergic way (see fourth and fifth diagrams) until an almost full rotation is achieved. The bunching couldn't be achieved at these moderate energies unless the average beam energy wouldn't be raised by acceleration to damp the outward pressure of the longitudinal space charge forces, but unfortunately the acceleration process induces a phase space curvature that in turns increases the longitudinal emittance and prevents reaching much shorter bunch lengths. If the process is pushed up to its limit the beam reaches the exit of the compressor (sixth diagram, average energy 24 MeV) being over-compressed. This means that the energy spread is now no longer correlated and it could not be corrected by a further acceleration off-crest to higher energies – as is typically done in Linacs with magnetic compressors. We will show however in the next section an example of partial compression with further energy spread correction, that is possible by just adjusting the injection phase into the compressor.

As discussed, the crucial feature of this method is the use of a slow RF structure such that the beam can be injected at a speed slower than that of the wave but it leaves the structure, after acceleration, being faster than the wave itself, i.e. at energies larger than the resonant gamma. The phase slippage is quite large during the compression, even up to 50-60° RF.

It would be possible to achieve such a phase slippage even in standard TW structures which are non-detuned, i.e. those having a wave phase velocity equal to the speed of light, for which the resonant gamma is actually infinity. In this structures the beam would be always slower than the wave, so that it would not be possible not even to reach the resonant gamma by the end of the RF compressor.

We anyway checked the potentialities of RF compressors based on non-detuned structures just by running Homdyn on the LCLS injector scheme and using low injection phases in the first TW structure (i.e. we repeated the simulations presented in Fig.4-5 just by setting $\Delta k = 0$). The maximum peak current achievable is in this case 250 A at 28 MeV exit energy from the compressor, with a longitudinal emittance of 150 keV·mm.

RF RECTILINEAR COMPRESSORS WITH HIGHER HARMONIC CORRECTORS

Since the main limitation to the performances of the RF compressor basically comes from the RF nonlinearities, a possible solution to this is the use of higher harmonics superimposed to the main RF field, aimed at correcting locally the RF curvature and avoid the huge increase of longitudinal emittance which is, in turn, responsible of the limitation in the minimum bunch length achievable.

The superposition of higher harmonics is indeed not conceivable inside the same slow wave structure of the compressor: it couldn't even be placed after the compression if this is pushed up to the overcompression limit (as shown in Fig.8), hence it's natural to conceive a lay-out in which the higher harmonic structure is placed before the compressor, i.e. in the drift between the gun and the compressor itself. This drift (about 1 m long) is needed to accomplish the first part of the transverse emittance correction by properly focusing the beam plasma oscillations through it.

We therefore consider here a lay-out as shown in Fig.9, with a short (7 cell) third harmonic structure placed right before the RF compressor, which is now followed by 5 standard 3 m long S-band SLAC TW structures in order to raise the beam energy up to 500 MeV, where the transverse space charge effects are expected to be negligible even for a 2 kA beam from the point of view of the emittance correction budget.

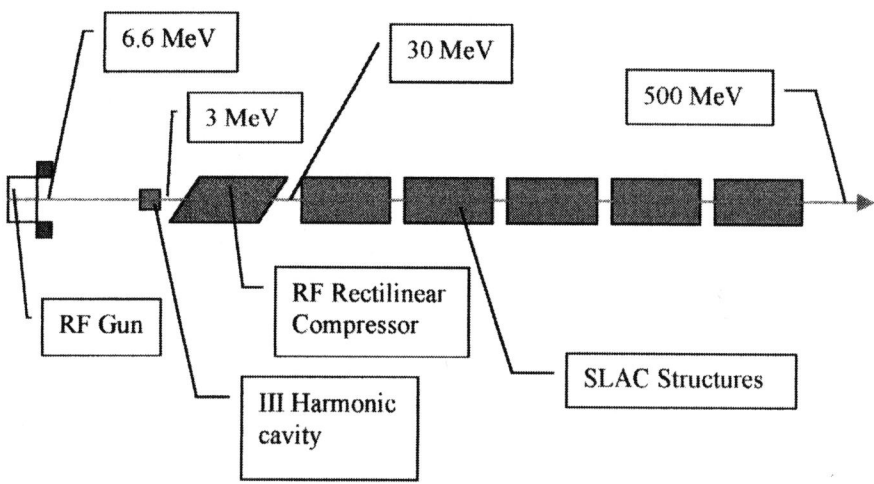

FIGURE 9. Layout of a 500 MeV Linac based on a RF compressor with 3rd harmonic correction for the generation of a 2.3 kA beam with no use of magnetic compressors.

The RF compressor is set at same conditions as in the previous example, i.e. $\gamma_r = 27$ ($\Delta k = 0.07\%$), driven at 23 MV/m accelerating gradient, while the following structures are driven at 27 MV/m (but the first one, set at 31 MV/m).

The 3rd harmonic structure, being 8.2 cm long, is driven at 40 MV/m at a frequency of 8568 MHz (no detuning). Its role is indeed to pre-correct the RF induced curvature in the longitudinal phase space distribution as that shown in Fig.8: in order to achieve this task is actually enough that the total energy decrease induced by the 3rd harmonic structure is close to 1/9 of the total energy increase produced by acceleration into the RF compressor. This would in fact correct at least the second order derivative (scaling like n^2, n being the harmonic number) of the phase space curvature, with a significant benefit to the longitudinal emittance.

This rule of the thumb would set a 3rd harmonic total energy decrease at 2.6 MeV, just one ninth of the total energy gain in the compressor, i.e. 23.5 MeV. From the simulations we found instead that the best results are obtained for an energy decrease of about 3.5 MeV, that corresponds to 40 MV/m acc. gradient. The reason for this disagreement is that the beam enters the compressor at 3 MeV, hence the energy gain and phase slippage are higher than before – as for the compression factor.

We should also notice that the exit energy from the compressor is in this case higher (30 MeV) than the case presented in previous section (see Fig.4): the reason is the higher phase slippage toward the RF field crest due to a lower injection energy in the compressor (3 MeV instead of 6.5 MeV).

Eventually the performances of the system are dramatically improved, thanks to the longitudinal emittance correction performed by the 3rd harmonic field, which brings it down to 25 keV·mm at the exit of the compressor, as opposed to 155 keV·mm without the 3rd harmonic correction. The beam current and energy throughout the Linac are plotted in Fig.10, while the energy spread and longitudinal emittance in Fig.11.

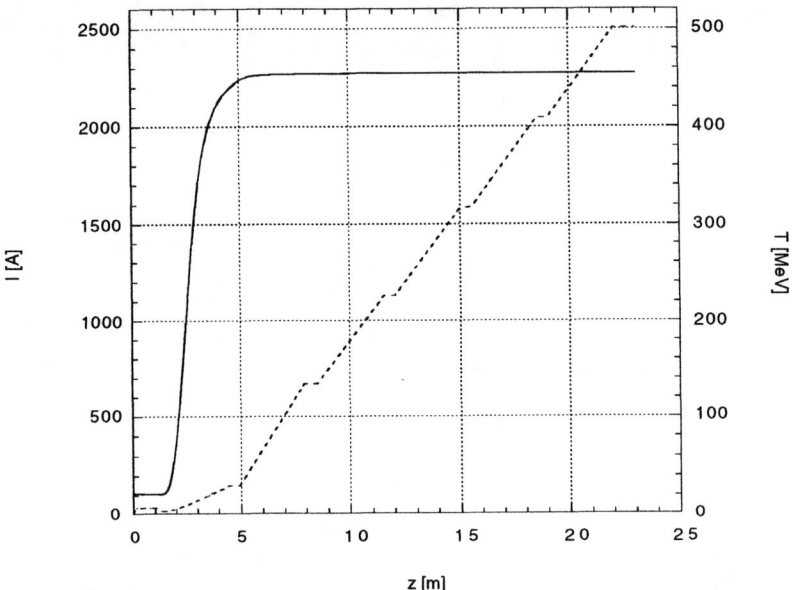

FIGURE 10. Beam current (solid line, left scale, in A) and energy evolution (dashed line, right scale, in MeV) through a 500 MeV Linac with implementation of a RF rectilinear compressor and a 3rd harmonic corrector.

The final peak current value is 2.3 kA at full compression, with an energy spread of 0.13 % at the end of the Linac, and a normalized rms transverse emittance still at 0.6 mm·mrad. It's interesting noticing the longitudinal emittance correction mechanism developed by the 3rd harmonic structure, located at 1.2 m from the photo-cathode: it actually blows up the longitudinal emittance from 15 to 37 keV·mm, as plotted in Fig.11, because of the positive curvature induced to the longitudinal phase space distribution, which is however corrected further on by the negative curvature contribution produced by the main RF field in the compressor. This brings eventually back the longitudinal emittance to its final value of 24 keV·mm, which allows for a compression factor almost 3 times larger than that obtained in absence of the 3rd harmonic corrector (see Fig.4).

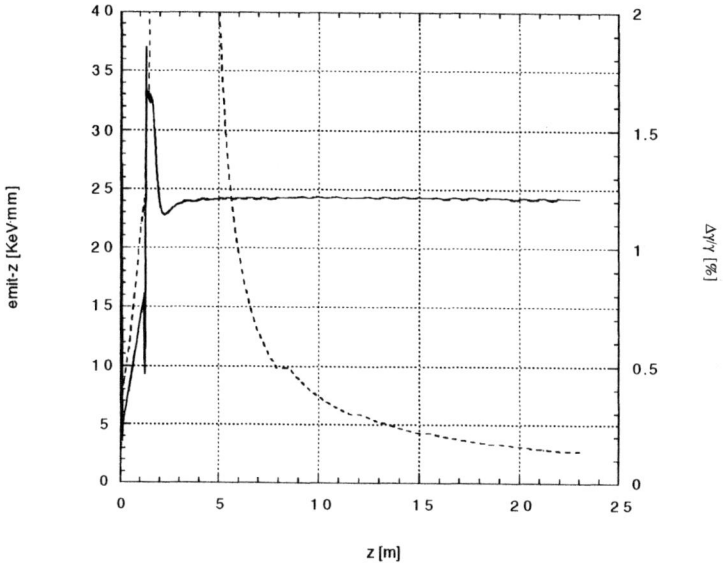

FIGURE 11. . Energy spread in % (dotted line, right scale) and longitudinal emittance in keV mm (solid line, left scale) evolution as for Fig.10 .

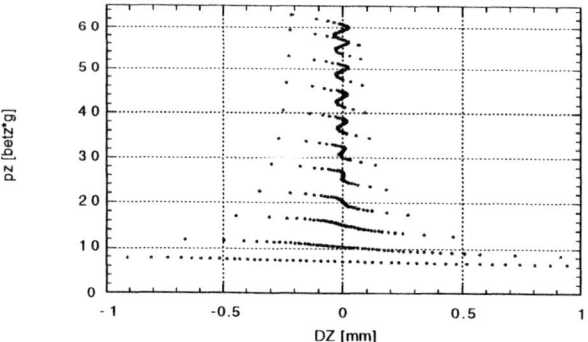

FIGURE 12. . Longitudinal phase space distributions taken at different times throughout the compressor for the case of total compression.

The compression is pushed in this case up to its maximum limit, as shown in Fig.12, where the phase space distributions are plotted at various times throughout the compressor. The residual energy spread is totally uncorrelated to the phase, hence not recoverable any longer.

If the compression is performed partially so that a linear correlation is still present at the exit of the compressor between the energy and phase of particles in the bunch, this correlated energy spread can be corrected by running the beam slightly off-crest in the following accelerating sections: as an example, it is possible to produce a beam with peak current of 600 A and a final energy spread (at the Linac exit, 500 MeV) of 0.04 % . The evolution along the Linac of the corresponding phase space distributions is reported in Fig.13

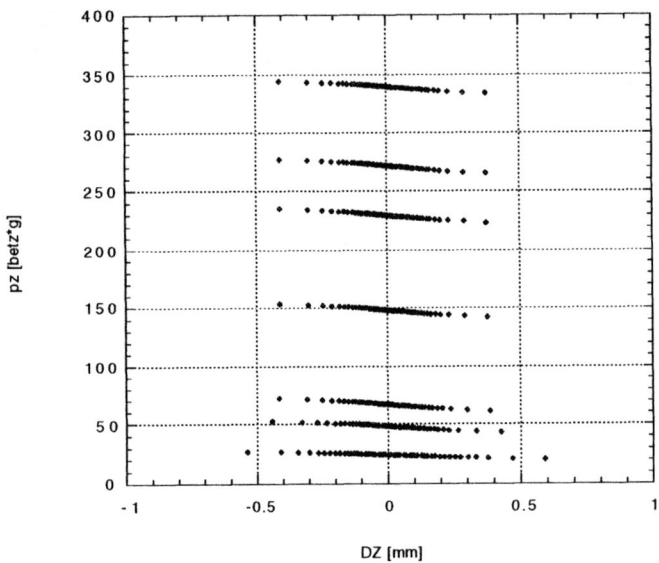

FIGURE 13. . Longitudinal phase space distributions taken at different times throughout the Linac for the case of partial compression.

The transverse dynamics is shown in Fig.14, where the rms beam envelope and normalized emittance are plotted along the Linac: the final emittance is still close to 0.55 mm·mrad, and this is again achieved by a careful setting of the field amplitudes and the locations of many solenoids placed around the RF compressor and the following two accelerating sections in order to keep focused the beam despite its increasing peak current. To this purpose, the RF compressor had to be placed closer to the RF gun exit, i.e. at 1.4 m far from the photocathode instead of 1.5 as for the LCLS nominal layout. Details of the transverse dynamics in this region are plotted in the lower diagram of Fig.14. No real systematic analysis has been done to optimize the transverse emittance up to 500 MeV: adding more solenoids in the last two accelerating sections would definitely help in taking the beam envelope better under control and getting a better emittance at the Linac exit.

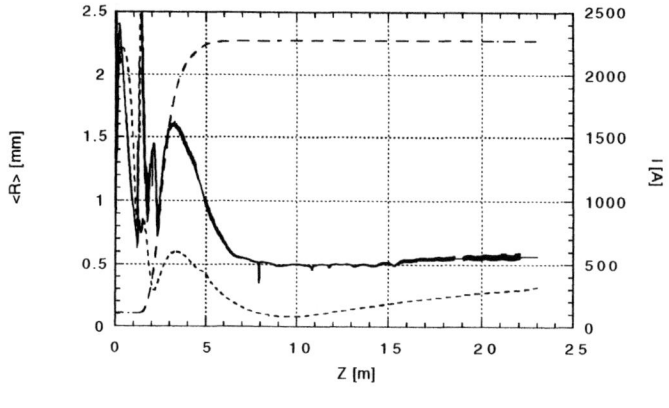

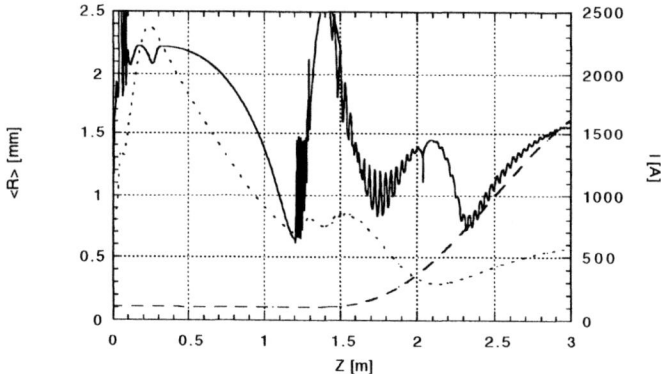

FIGURE 14. Rms beam envelope in mm (dotted line, left scale), rms normalized transverse emittance in mm mrad (solid line, right scale) and beam peak current in A (solid line, right scale) along the Linac. The lower diagram shows the details of the upper one along the first 3 m of RF gun, drift, 3rd harmonic structure (from z=1.2 to z=1.28) and RF compressor (from z=1.4 to z=4.4)

RF RECTILINEAR COMPRESSORS WITH STANDING WAVE STRUCTURES

The analysis performed so far has been focused on the use of traveling wave structures, where the analytical treatment reported in the introduction strictly applies.

There is however a relevant interest also in applying this concept in the domain of Linacs based on standing wave structures. For this case an analytical theory of the rectilinear compressor has not yet been investigated in details. Nevertheless we expect to apply the previous treatment provided that only the forward travelling wave component of the standing wave is considered.

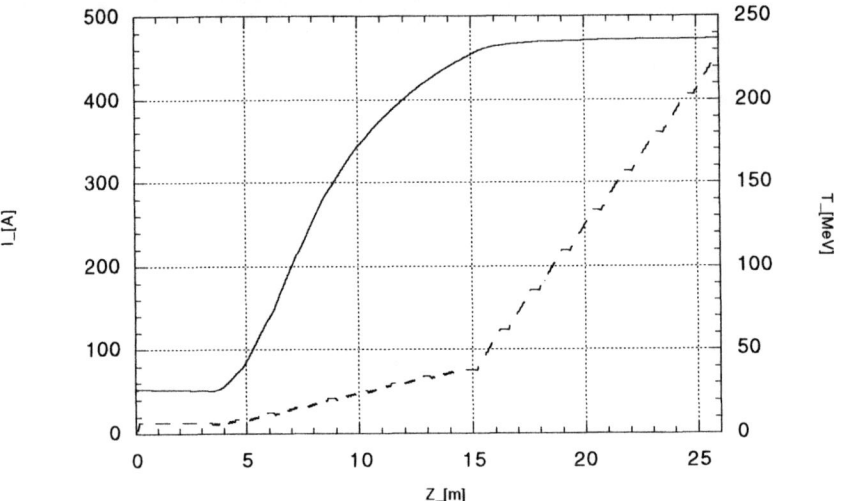

FIGURE 15. Beam current (solid line, left scale) and energy evolution (dashed line, right scale) through the TESLA-FEL injector with implementation of a 11 m long RF compressor (first cryomodule).

We have applied for demonstration the rectilinear compressor concept to an injector beam line based on superconducting standing wave structures, as the one designed for the TESLA FEL project [9]. We have considered in this case a 1.3 GHz rf gun, 1.5 cells long, followed by two cryomodules each one containing 8, 1.3 GHz, 9 cells standing wave structures. A 1nC bunch is extracted from the cathode by a 20 ps long laser pulse in a 60 MV/m rf peak field and accelerated up to 6.6 MeV (γ_0=14) at the gun exit, resulting in a 50 A peak current (with 0.7 % energy spread and 0.6 mm-mrad rms transverse normalized emittance minimum). By detuning the first group of 8 cavities so to increase the wave number by 0.08 % , corresponding to a resonant gamma γ_r=25.5, and operating the detuned structures at 50 MV/m peak field, corresponding to α=1.8 for the forward travelling wave component, the expected compression factor results to be C=4.

The effectiveness of the rectilinear compressor scheme based on standing wave structures is clearly visible in Fig. 15 where the energy gain and the peak current are reported up to the end of the second cryomodule. A compression ratio higher than 9 has been obtained without any attempt to optimize the longitudinal and transverse emittances nor the energy spread of the beam.

We can definitely speculate, nevertheless, that even in SW Linacs it should be possible to attain peak currents in excess of 1 kA by implementation of higher harmonic correctors as it was shown in previous section.

CONCLUSIONS

On the basis of the very preliminary analysis presented in this paper we may conclude that the use of RF rectilinear compressor can be very helpful in the design of SASE X-ray FEL drivers, either in alternative to magnetic compressors, or to alleviate the criticality of these devices, i.e. in order to avoid multi-staged magnetic compressors that appear critical in those projects.

In fact, we have shown here that peak currents in excess of 2 kA are achievable with the use of properly designed RF compressors. Since there is no systematic study performed so far on these devices we might speculate that one should be able to reach even higher current levels (> 3 kA ?). In case one is aiming at higher currents the use of a magnetic compressor could be unavoidable, but its criticality should be diminished by the lower compression factor required to it – because the beam would already have more than 1 kA of peak current at its entrance.

In order to summarize all the different examples presented in this paper, we listed in Table 1 the different options available: this enlighten the flexibility of this method that can be applied in several different situations, from Linacs based on TW room temperature structures to Superconducting SW Linacs, at almost any frequency of interest in these designs (included plasma accelerators, where the method was firstly conceived).

TABLE 1. Performances of RF Rectilinear Compressors.

Type of structure	Peak Current [A]	Energy spread [%] @ 500 MeV	Longit. emittance [KeV mm]
zero detuning @ S-band TW	250	0.1	150
0.07% detuning @ S-band TW	870	0.3	155
plus 3rd Harmonic	2300	0.13	25
plus 3rd Harmonic – partial compression	600	0.04	25
0.08% detuning @ L-band SW	470	not optimized	not optimized

ACKNOWLEDGMENTS

We acknowledge very helpful discussions with G.D'Auria and C.Rossi on slow-wave structures for the RF compressor.

REFERENCES

1. M. Ferrario, T. C. Katsouleas, L. Serafini, I. Ben-Zvi.,"Adiabatic Plasma Buncher", in publication on *IEEE Trans. on Plasma Science*, **28**, no.4 (2000).
2. C. Clayton and L. Serafini, *IEEE Trans. on Plasma Science* **24**, p.400 (1996).
3. C.S. Liu, V. K. Tripathi, *Interaction of Electromagnetic Waves with Electron Beams and Plasmas* , World Scientific Publ., 1994, pp. 60-63.
4. J. Arthur et al., *Linac Coherent Light Source (LCLS) Desing Study Report*, SLAC-R-251, Apr. 1998.
5. M. Ferrario, J. Clendenin, D. Palmer, J. Rosenzweig, L.Serafini,, "HOMDYN study for the LCLS photoinjector", in publication on *Proc. ICFA Workshop on The Physics of High Brightness Beams*, Ed. J. Rosenzweig and L. Serafini, World Scientific Publ., (2000).
6. M. Ferrario, A. Mosnier, L. Serafini, F. Tazzioli, J.M. Tessier, *Particle Accelerators* **52**, p.1 , (1996).
7. L. Serafini and J. B. Rosenzweig, *Physical Review* **E 55**, p.7565 (1997).
8. D. H. Dowell, T. D. Hayward, A. M. Vetter, *IEEE Cat.* no. 95CH35843, p.992 (1996).
9. Conceptual Design of a 500 GeV e+e- Linear Collider with Integrated X-ray Laser Facility, Ed. R. Brinkmann, G. Materlik, J. Rossbach, A. Wagner, DESY-1997-048.
10. The analytical model takes indeed into account only initial energy and phase spreads which are uncorrelated and uniformly distributed around the average injection energy and phase. The simulations are instead done with injection conditions for the beam (see fig.8) that imply correlated longitudinal distributions with only partial fill-up of the longitudinal phase space: this may be the reason why the analytical model tend to underestimate the compression factor.

Effects of Statistical Roughness on the Propagation of Electromagnetic Fields in a Circular Waveguide

E. Di Liberto*, F. Frezza†, L. Palumbo‡

*Università di Roma "La Sapienza"
†Dip. di Ingegneria Elettronica-Università di Roma "La Sapienza"
‡Dip. Energetica-Università di Roma "La Sapienza"

Abstract. The Green's function method (GFM) is used to describe the effects of a statistical roughness on the propagation of electromagnetic fields in a circular waveguide. The GFM method, applied to a scalar Helmholtz equation with perturbed boundary conditions, allows us to treat multiple scattering of the field. Theoretical results for azimuthal and longitudinal roughness are reported, and the phase velocity of the field is discussed in both cases.

INTRODUCTION

In the design of future Free Electron Lasers and Linacs very short bunches of high intensity, small emittance and small energy spread are planned to be used. It has been recently remarked [1] that even the roughness of beam pipe surface might excite parasitic fields increasing emittance and energy spread.

Several simplified models have been developed to evaluate these effects, assuming irregularities either as small bumps with simple shape and random distribution on a smooth surface [1] or as a periodic corrugation of beam tube [2]; a further model [1] describes the roughness, into a local cartesian reference frame, as a random function $y = h(x, z)$ of surface coordinates x and z and longitudinal impedance is calculated depending on statistical properties of the process h, assuming small local derivatives (small-angle approximation, $|\nabla h| \ll 1$).

In this paper we investigate the effects of surface roughness on an electromagnetic wave travelling in a circular waveguide whose boundary is given by the 2-D equation $a + \tilde{\rho}(\varphi, z)$.

It should be noted that an electromagnetic wave incident on a rough surface is subjected to diffraction phenomena which can be divided into two groups according to the number of total reflections: the first group comprises processes like incidence on sea surface or irregular ground, where the wave undergoes a single act of scattering and only slight distorsions of the field are produced [3]; the second group

CP581, *Physics of, and Science with, the X-Ray Free-Electron Laser*, edited by S. Chattopadhyay et al.
© 2001 American Institute of Physics 0-7354-0022-9/01/$18.00

is made by the phenomena that take place in domains bounded by two or more scattering surfaces, like in rough resonant cavities or waveguides, where a field undergoes multiple scattering along propagation path and a simple pertubation theory is not very fruitful to describe it correctly.

The GFM allows us to treat distributed summation of waves propagating in a waveguide with randomly rough shape, approximating an homogeneous boundary condition on the perturbed surface S_p by an inhomogeneous condition calculated on the average, i.e. ideal, surface S (see Fig. 1). Being $\tilde{\rho}(\varphi, z)$ a random function of transverse coordinates, we calculate, depending on its statistical features (variance and correlation function), the phase velocity shift for the electromagnetic field.

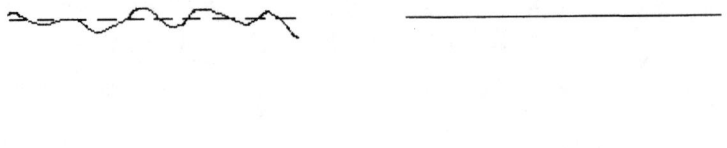

$$S_p \qquad\qquad\qquad\qquad S$$

FIGURE 1. Perturbed and average surface.

THE GREEN'S FUNCTION METHOD

General Description

Let us consider the function $G(\mathbf{r}, \mathbf{r}')$ solution of three dimensional Helmholtz equation

$$\nabla^2 G(\mathbf{r}, \mathbf{r}') + k^2 G(\mathbf{r}, \mathbf{r}') = \delta(\mathbf{r} - \mathbf{r}') \tag{1}$$

with the mixed boundary condition (BC)

$$\frac{\partial G(\mathbf{r}, \mathbf{r}')}{\partial n_p} + \eta_o\, G(\mathbf{r}, \mathbf{r}')\Big|_{\mathbf{r} \in S_p} = 0 \tag{2}$$

where η_o is an integro-differential operator and $\mathbf{n}_p$ is a unit vector normal to the surface S_p, described by the radius vector (see Fig. 2)

$$\mathbf{r} = \mathbf{r}_s + \mathbf{n}(\mathbf{r}_s)\,\zeta(\mathbf{r}_s) \tag{3}$$

where $\mathbf{r}_s$ lies on the ideal surface S that undergoes shape deviations given by the random function $\zeta(\mathbf{r}_s)$. In the following we will consider as known the function $G_o(\mathbf{r}, \mathbf{r}')$, that is the solution of eq. (1) when no perturbation is present.

Assuming that the *small angle approximation*

$$\zeta(\mathbf{r}_s) \ll a \qquad |\nabla_t \zeta(\mathbf{r}_s)| \ll 1, \qquad \nabla_t = \nabla - \mathbf{n}(\nabla \cdot \mathbf{n})$$

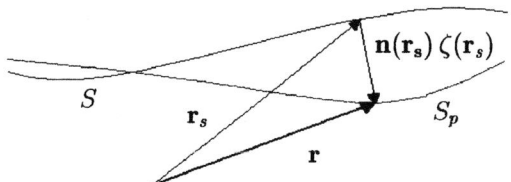

FIGURE 2. Radius vector.

is verified, and substituting (3) into (2), the BC can be transferred on the ideal surface S by means of an expansion in the perturbation parameter ζ, retaining only first order terms

$$\frac{\partial G}{\partial n} + \eta_o\, G - \nabla_t\zeta \cdot \nabla_t G + \zeta \frac{\partial^2 G}{\partial n^2} + \eta_o\zeta \frac{\partial G}{\partial n}\bigg|_{\mathbf{r}\in S} = 0 \qquad (4)$$

This equation is known as approximated or *effective* boundary condition (EBC) that can be shortly written as

$$\frac{\partial G}{\partial n} + \eta_o G + \hat{V} G\ \bigg|_{S} = 0 \qquad (5)$$

that is an unperturbed condition corrected by a perturbation term proportional to the random parameter ζ. Using Green's theorem it is possible to express $G(\mathbf{r}, \mathbf{r}')$ depending on $G_o(\mathbf{r}, \mathbf{r}')$, provided that their respective BCs are introduced into the surface integral term, as follows:

$$\begin{aligned} G(\mathbf{r}, \mathbf{r}') &= G_o(\mathbf{r}, \mathbf{r}') + \int_S \left[\frac{\partial G_o}{\partial n} G + G_o \frac{\partial G}{\partial n}\right] dS_s = \\ &= G_o(\mathbf{r}, \mathbf{r}') + \int_S G_o(\mathbf{r}, \mathbf{r}_s)\hat{V}(\mathbf{r}_s)G(\mathbf{r}_s, \mathbf{r}')dS_s \end{aligned} \qquad (6)$$

The integral equation (6) shows a functional dependence of $G(\mathbf{r}, \mathbf{r}')$ on the random parameter $\hat{V}(\zeta)$ and makes necessary its description as a statistical process. As regards to the *average* Green's function $\mathcal{G}(\mathbf{r}, \mathbf{r}') = \,<G(\mathbf{r}, \mathbf{r}')>$, i.e. the first order statistical moment, it can be shown that it is still solution of an integral equation that is known as *Dyson equation* [5]:

$$\mathcal{G}(\mathbf{r}, \mathbf{r}') = G_o(\mathbf{r}, \mathbf{r}') + \iint_S G_o(\mathbf{r}, \mathbf{r}_{s_1})\, \mathcal{M}(\mathbf{r}_{s_1}, \mathbf{r}_{s_2})\, \mathcal{G}(\mathbf{r}_{s_2}, \mathbf{r}')\, dS_{s_1} dS_{s_2}, \quad \mathbf{r}_{s_i} \in S \quad (7)$$

The *mass operator* function $\mathcal{M}(\mathbf{r}_{s_1}, \mathbf{r}_{s_2})$ introduces the dependence on the random function $\zeta(\mathbf{r}_s)$ and can be expressed as an infinite series, performing a multiple iteration[1] of eq. (7) [5,6]. If one let $\mathbf{r}$ approach S, because of the properties of unperturbed Green's function[2] $,\mathcal{G}(\mathbf{r},\mathbf{r}')$ verifies a *non-local* BC given by

$$\left(\frac{\partial}{\partial n} + \eta_o\right)\mathcal{G}(\mathbf{r},\mathbf{r}')\bigg|_S = \int_S \mathcal{M}(\mathbf{r}_s,\mathbf{r}_{s_1})\mathcal{G}(\mathbf{r}_{s_1},\mathbf{r}')dS_{s_1} \tag{8}$$

Therefore, assuming to know the *unperturbed* Green's function $G_o(\mathbf{r},\mathbf{r}')$, solution of the Helmholtz's problem (HP)

$$\begin{cases} \nabla^2 G_o(\mathbf{r},\mathbf{r}') + k^2\,G_o(\mathbf{r},\mathbf{r}') = \delta(\mathbf{r}-\mathbf{r}') \\ \left(\frac{\partial}{\partial n} + \eta_o\right)G_o\big|_{\mathbf{r}\in S} = 0 \end{cases}$$

and introducing random deviations in the boundary's shape, it is possible to bring back to a similar form even the *perturbed* equations for the Green's function $G(\mathbf{r},\mathbf{r}')$, by means of the approximation in a Taylor series of the homogeneous BC calculated on S_p with the inhomogeneous EBC verified on the ideal surface S:

$$\begin{cases} \nabla^2 G(\mathbf{r},\mathbf{r}') + k^2\,G(\mathbf{r},\mathbf{r}') = \delta(\mathbf{r}-\mathbf{r}') \\ \left(\frac{\partial}{\partial n} + \eta_o + \hat{V}\right)G\big|_{\mathbf{r}\in S} = 0 \end{cases}$$

For the *average* Green's function $\mathcal{G}(\mathbf{r},\mathbf{r}')$, solution itself of the Helmholtz equation[3], a non-local BC has to be verified on the ideal surface S and the HP equivalent to eq. (7) appears as:

$$\begin{cases} \nabla^2\mathcal{G}(\mathbf{r},\mathbf{r}') + k^2\,\mathcal{G}(\mathbf{r},\mathbf{r}') = \delta(\mathbf{r}-\mathbf{r}') \\ \left(\frac{\partial}{\partial n} + \eta_o\right)\mathcal{G}\big|_S = \int_S \mathcal{M}(\mathbf{r}_s,\mathbf{r}'_s)\,\mathcal{G}(\mathbf{r}'_s,\mathbf{r}')\,dS \end{cases}$$

Dirichlet Boundary condition

If the Green function verifies a Dirichlet boundary condition $G(\mathbf{r},\mathbf{r}') = 0$ on S_p, then the EBC on S becomes

$$G(\mathbf{r}_s,\mathbf{r}') + \zeta(\mathbf{r}_s)\frac{\partial G(\mathbf{r}_s,\mathbf{r}')}{\partial n} = 0 \qquad \mathbf{r}_s \in S \tag{9}$$

---

[1] Actually the GFM was originally introduced to describe wave propagation into 3-D random media, employing iteratively the Feynman diagram technique: the treatment shown in this paper derives from an extension to the case of boundary perturbations [5].
[2] In fact the boundary condition of the unperturbed Green's function G_o for $\mathbf{r} \to S$ should be written as

$$\frac{\partial G_o}{\partial n} + \eta_o\,G_o\bigg|_{\mathbf{r}\to S} = \delta(\mathbf{r}-\mathbf{r}')$$

that simply reduces to a form analogous to eq. (2), provided that the second argument of the delta function represents a point inside the bounded domain.
[3] In fact the operations of averaging and derivating are linear and thus interchangeable.

The integral equation derived from Green's theorem and verified by the perturbed Green's function appers as:

$$G(\mathbf{r}, \mathbf{r}') = G_o(\mathbf{r}, \mathbf{r}') - \int_S \frac{\partial G_o(\mathbf{r}, \mathbf{r}_s)}{\partial n} \zeta(\mathbf{r}_s) \frac{\partial G(\mathbf{r}_s, \mathbf{r}')}{\partial n} dS_s \qquad (10)$$

In this case the Dyson equation for the average Green's function $\mathcal{G}(\mathbf{r}, \mathbf{r}')$ appears as [4]:

$$\mathcal{G}(\mathbf{r}, \mathbf{r}') = G_o(\mathbf{r}, \mathbf{r}') + \iint_S \frac{\partial G_o(\mathbf{r}, \mathbf{r}_s')}{\partial n_s'} M(\mathbf{r}_s', \mathbf{r}_s'') \frac{\partial \mathcal{G}(\mathbf{r}_s'', \mathbf{r}')}{\partial n_s''} dS_s' dS_s'' \qquad (11)$$

Considering a first-order iterative solution for (10)-(11) (see footnote no. 1), we can derive an approximated expression for the mass operator $M(\mathbf{r}_{s_1}, \mathbf{r}_{s_2})$ of eq. (11), clearly introducing the statistical features of the random process $\zeta(\mathbf{r}_s)$ into the problem. As regards to the eq. (10) one obtains:

$$G(\mathbf{r}, \mathbf{r}') = G_o(\mathbf{r}, \mathbf{r}') - \int \frac{\partial G_o(\mathbf{r}, \mathbf{r}_s)}{\partial n_s} \zeta(\mathbf{r}_s) \frac{\partial G_o(\mathbf{r}_s, \mathbf{r}')}{\partial n_s} dS_s +$$
$$+ \iint_S \frac{\partial G_o(\mathbf{r}, \mathbf{r}_s')}{\partial n_s'} \zeta(\mathbf{r}_s') \frac{\partial^2 G_o(\mathbf{r}_s', \mathbf{r}_s'')}{\partial n_s' \partial n_s''} \zeta(\mathbf{r}_s'') \frac{\partial G_o(\mathbf{r}_s'', \mathbf{r}')}{\partial n_s''} dS_s' dS_s'' + \ldots \qquad (12a)$$

The physical meaning of the former equation is that the e.m. field propagating along the waveguide is the result of an infinity of contributions coming from equivalent surface sources radiating multiply-scattered waves. Averaging both sides of the previous equation, it results that:

$$\mathcal{G}(\mathbf{r}, \mathbf{r}') = G_o(\mathbf{r}, \mathbf{r}') - \int \frac{\partial G_o(\mathbf{r}, \mathbf{r}_s)}{\partial n_s} \underbrace{< \zeta(\mathbf{r}_s) >}_{=0} \frac{\partial G_o(\mathbf{r}_s, \mathbf{r}')}{\partial n_s} dS_s +$$
$$+ \iint_S \frac{\partial G_o(\mathbf{r}, \mathbf{r}_s')}{\partial n_s'} \left[\sigma^2(\mathbf{r}_s) W(\mathbf{r}_s', \mathbf{r}_s'') \frac{\partial^2 G_o(\mathbf{r}_s', \mathbf{r}_s'')}{\partial n_s' \partial n_s''} \right] \frac{\partial G_o(\mathbf{r}_s'', \mathbf{r}')}{\partial n_s''} dS_s' dS_s'' + \ldots \qquad (12b)$$

where $< \zeta(\mathbf{r}_s) > = 0$ represents the mean value of the random perturbation process, $\sigma^2(\mathbf{r}_s) = < \zeta^2(\mathbf{r}_s) >$ is its variance and $W(\mathbf{r}_s, \mathbf{r}_s') = \sigma^{-2}(\mathbf{r}_s) < \zeta(\mathbf{r}_s)\zeta(\mathbf{r}_s') >$ is its correlation coefficient. Performing an iteration of eq. (11) and making a comparison with eq. (12b), one can express the mass operator as a series:

$$M(\mathbf{r}_s, \mathbf{r}_s') = \sigma^2(\mathbf{r}_s) W(\mathbf{r}_s, \mathbf{r}_s') \frac{\partial^2 G_o(\mathbf{r}_s, \mathbf{r}_s')}{\partial n_s \partial n_s'} + \ldots \qquad (13)$$

The first term appears to be proportional to the ratio $(\sigma/a)^2$ while higher order ones result proportional to $(\sigma/a)^4$ and therefore will be neglected for the following, according to the so-called *Buorret approximation* [3].

CIRCULAR WAVEGUIDE WITH RANDOM ROUGHNESS

In a cylindrical system of coordinates $\{\rho, \varphi, z\}$ a set of solutions corresponding to $\text{TM}^{(z)}$ e.m. waves can always be derived from the non-null longitudinal component of electric field E_z, considered as potential function, that is solution of

$$\nabla^2 E_z + k^2 E_z = 0 \tag{14}$$

with the further condition $\mathbf{n} \times \mathbf{E}|_S = 0$ when a bounding metallic surface S is present. If this surface can be described as $a + \tilde{\rho}(\varphi, z)$, then its unit vector is $\mathbf{n}_p \cong [1, -\frac{\partial \tilde{\rho}}{\partial \varphi}, -\frac{\partial \tilde{\rho}}{\partial z}]$ and the vanishing condition for $\mathbf{E}$ becomes

$$\frac{\partial \tilde{\rho}}{\partial z} E_\rho + E_z = 0 \tag{15}$$

Performing separated analysis for azimuthal and longitudinal corrugations, we have made differences stand out as regards eigenvalue spectrum, pointing out typical features of both cases.

Azimuthal Roughness

If the boundary surface of the waveguide is given by $a + \tilde{\rho}(\varphi)$, simple geometrical considerations allow us to treat a bidimensional Helmholtz equation in the transverse plane, assuming for all solutions an exponential dependence[4] on z like $e^{-jk_z z}$. Since $\tilde{\rho}$ varies only with azimuthal coordinate, equation (15) entails $E_z = 0$ on S_p. The technique presented in previous section has been applied to a Green's problem equivalent to equations (14)-(15).
The EBC for azimuthal perturbations appears as

$$G(\rho, \varphi, \rho', \varphi') + \frac{\partial G(\rho, \varphi, \rho', \varphi')}{\partial \rho} \left. \tilde{\rho}(\varphi) \right|_{\rho=a} = 0 \tag{16}$$

and the Green's theorem for bidimensional domains leads to the integral Dyson equation

$$\mathcal{G}(\mathbf{r}_t, \mathbf{r}_t') = G_o(\mathbf{r}_t, \mathbf{r}_t') + \iint\limits_S \frac{\partial G_o(\mathbf{r}_t, \mathbf{r}_{t_s})}{\partial n_s} M(\mathbf{r}_{t_s}, \mathbf{r}_{t_s'}) \frac{\partial \mathcal{G}(\mathbf{r}_{t_s}, \mathbf{r}_t')}{\partial n_s'} d\mathbf{r}_{t_s} d\mathbf{r}_{t_s}' \tag{17}$$

where $\mathbf{r}_{t_s} = (a, \varphi)$.
In this situation the non-local boundary condition is

$$\left. \mathcal{G}(a, \rho', \varphi, \varphi') \right|_{\rho=a} = \int_S M(\varphi, \varphi'') \left. \frac{\partial \mathcal{G}(\rho'', \varphi'', \rho', \varphi')}{\partial \rho''} \right|_{\rho''=a} a \, d\varphi'' \tag{18}$$

---

[4] Let us remind that this is actually equivalent to perform a Fourier transform.

because of the properties of unperturbed Green's function for Dirichlet conditions. If one assumes $\tilde{\rho}(\varphi)$ as a statistically uniform process this means that

$$\sigma^2(\varphi) = \sigma^2 \qquad\qquad W(\varphi, \varphi') = W(\varphi - \varphi')$$

Physically this assumption makes the problem uneffected by any waveguide rotation and the field excited in the point $P[\varphi]$ by an impulsive source located in the point $P'[\varphi']$ (i.e. the Green's function) becomes only dependent on $[\varphi - \varphi']$ (see Fig. 3).

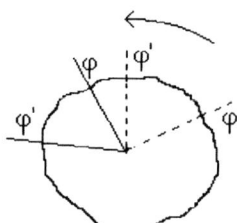

FIGURE 3. Stationary condition for azimuthal roughness.

Taking into account this property and developing both members of (18) in Fourier series, a mixed boundary condition at $\rho = a$ emerges for harmonic components $\mathcal{G}_n(\rho, \rho')$ of average Green function

$$\mathcal{G}_n(a, \rho') = 2\pi a\, M_n\, \left.\frac{\partial \mathcal{G}_n(\rho, \rho')}{\partial \rho}\right|_{\rho=a} \qquad (19)$$

Being $\mathcal{G}_n(\rho, \rho')$ solutions of a Bessel differential equation[5] verifying condition (19), their expression is given by [6]

$$\mathcal{G}_n(\rho, \rho', k_z) = \frac{1}{Den(a, k_z, k_\rho)} \cdot \begin{cases} \psi_1(\rho)\Psi_2(\rho') & \rho \le \rho' \\ \psi_1(\rho)\Psi_2(\rho) & \rho \ge \rho' \end{cases} \qquad (20)$$

where

$$\begin{cases} \psi_1(\rho, n, k_z) = J_n(k_\rho\rho) \\ \psi_2(\rho, n, k_z) = J_n(k_\rho\rho) - [J_n(k_\rho a)/Y_n(k_\rho a)]\, Y_n(k_\rho\rho) \\ \Psi_2(\rho, n, k_z) = [\psi_1(a) + 2\pi a k_\rho M_n(k_z)\psi_1'(a)]\, \psi_2(\rho) - [2\pi a k_\rho M_n(k_z)\psi_2'(a)]\, \psi_1(\rho) \end{cases}$$

Since the singularities of the average Green's function coincide with the eigenvalues of the waveguide, then the perturbed dispersion equation appears in the form

$$J_n(k_\rho a) - 2\pi a k_\rho M_n(k_z) J_n'(k_\rho a) = 0 \qquad (21)$$

---

[5] See footnote no. 3.

An approximated evaluation of transverse eigenvalue shift is possible, assuming $|\delta k_\rho|$ much smaller than the distance between two adjacent roots and thus solving (21) as a perturbation of $J_n(k_\rho a) = 0$, finally obtaining

$$\delta k_\rho \cong 2\pi \, k_\rho \, M_n(k_z) \tag{22}$$

The random properties of the process $\tilde{\rho}$ appears by means of the Fourier components M_n of the mass operator $M(\varphi, \varphi')$ that is given (see eq. (13)) by

$$M(\varphi, \varphi') \approx \sigma^2 W(\varphi - \varphi') \left. \frac{\partial^2 G_o(\rho, \varphi, \rho', \varphi')}{\partial \rho \partial \rho'} \right|_{\rho = \rho' = a} + \dots \tag{23}$$

Substituting (23) into (22) it follows

$$\delta k_{\rho_{n,m}} = 2\pi k_{\rho_{n,m}} \frac{\sigma^2}{\pi a^4} \sum_{p=0}^{+\infty} \sum_{q=1}^{+\infty} \frac{\epsilon_p}{2} \xi_{p,q}^2 \frac{W_{n-p} + W_{n+p}}{k_{z_{p,q}}^2 - k_{z_{n,m}}^2} \tag{24}$$

where ϵ_p is the Neumann symbol, $\xi_{p,q}$ is the q-th zero of Bessel function $J_p(x)$, $k_{z_{p,q}} = [k^2 - (\xi_{p,q}/a)^2]^{1/2}$, and W_i represents the i-th Fourier harmonic of the correlation coefficient $W(\varphi, \varphi')$. In eq. (24) the term with index $\{n, m\}$ shows a null denominator as consequence of previous approximations. The resonance can be avoided performing a more accurate evaluation that finally leads, retaining only dominant term, to

$$\delta k_{z_{n,m}} = \left(\frac{dk_z}{dk_\rho} \right) \delta k_{\rho_{n,m}} = \left(\frac{\sigma}{a} \right) \frac{k_{\rho_{n,m}}^2}{k_{z_{n,m}}} \sqrt{W_o} \tag{25}$$

as regards propagation constant $k_z = [k^2 - k_\rho^2]^{1/2}$.

In Fig. 4 the normalized shift for $TM_{0,1}$ is plotted versus normalized frequency, for different values of kl, where l is the correlation length, assuming for $\tilde{\rho}(\varphi)$ a gaussian statistics:

$$W(\varphi, \varphi') = e^{-\frac{2a^2[1 - \cos(\varphi - \varphi')]}{l^2}} \tag{26}$$

The decreasing of phase velocity determined by eq. (25) appears to be very small and, within the range of validity of our approximations, no synchronuos field seems to be excited.

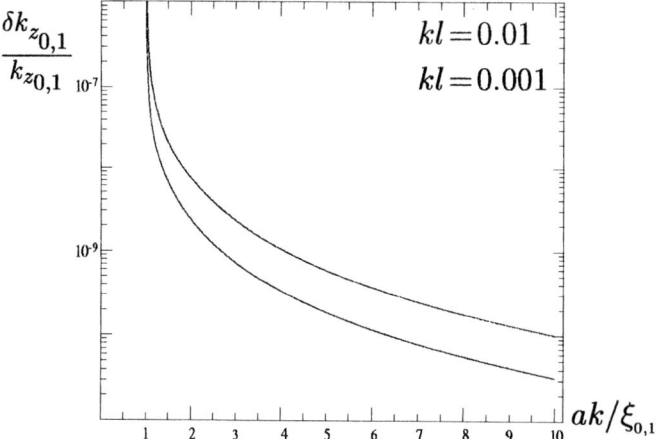

FIGURE 4. Normalized shift vs. normalized frequencies for azimuthal roughness.

Longitudinal Roughness

If the boundary surface of the waveguide is now given by $a + \tilde{\rho}(z)$, it is no longer possible to separate longitudinal propagation from transverse resonance: anyway one can still consider a bidimensional problem, now in the $\{\rho, z\}$ plane, with the condition $E_z = 0$ on the rough boundary, provided that the first term of eq. (15) might be neglectable in case of small perturbations with small gradient. Calculations absolutely similar to those of previous section again lead to the perturbed dispersion equation (21). The difference is in the form of mass operator transform, appearing now as a continuous convolution

$$M_n(k_z) = \frac{\sigma^2}{4\pi^2} \frac{1}{\pi a^4} \sum_{q=1}^{+\infty} \xi_{n,q}^2 \int_{-\infty}^{+\infty} \frac{\tilde{W}(k_z - k_z')}{k_{z_{n,q}}^2 - k_z'^2} \, dk_z' \qquad (27)$$

where $\tilde{W}(k_z)$ is the Fourier transform of the correlation coefficient $W(z - z')$. In order to determine eigenvalue shift, it is necessary to utilize the radiation principle [5] in the solution of the intergral present in eq. (27). It appears that now $\delta k_{z_{n,m}}$ holds real and imaginary part

$$\Re[\delta k_{z_{n,m}}] = \frac{k_{\rho_{n,m}}^2}{k_{z_{n,m}}} \frac{\sigma^2}{2\pi a^4} \sum_{q=Q(n)+1}^{+\infty} \xi_{n,q}^2 \frac{\tilde{W}(k_{z_{n,m}} + \jmath\alpha_{z_{n,q}})}{\alpha_{z_{n,q}}} \qquad (28a)$$

$$\Im[\delta k_{z_{n,m}}] = -\frac{k_{\rho_{n,m}}^2}{k_{z_{n,m}}} \frac{\sigma^2}{2\pi a^4} \sum_{q=1}^{Q(n)} \xi_{n,q}^2 \frac{[\tilde{W}(k_{z_{n,m}} - k_{z_{n,q}}) + \tilde{W}(k_{z_{n,m}} + k_{z_{n,q}})]}{\beta_{z_{n,q}}} \qquad (28b)$$

where Q verifies $\xi_{n,1} < ak < \xi_{n+1,1}$ and $\xi_{n,Q} < ak < \xi_{n,Q+1}$.

The incoherent scattering into other modes that can propagate *above cut-off* along the waveguide (both in forward and in backward direction, as shown in Fig. 5) causes an attenuation given by eq. (28 b), while the excited modes *below cut-off* decrease the phase velocity according to eq. (28 a).

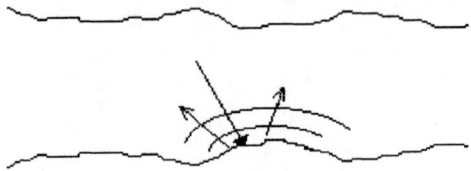

FIGURE 5. Scattered waves both in forward and backward direction.

In Fig. 6 the normalized attenuation for $TM_{0,1}$ is plotted versus normalized frequency, assuming for $\tilde{\rho}(z)$ a gaussian statistics

$$W(z - z') = e^{-\frac{(z-z')^2}{l^2}} \tag{29}$$

The unlimited resonance peaks corresponding to the different cut-off frequencies of the circular waveguide are a conseguence of several approximations formerly done, but anyway they are the trace of a real physical phenomenon: in fact an eigenwave propagating along the waveguide with a steep angle of incidence, being frequently reflected, is more influenced by roughness and thus the effect of perturbation is amplified.

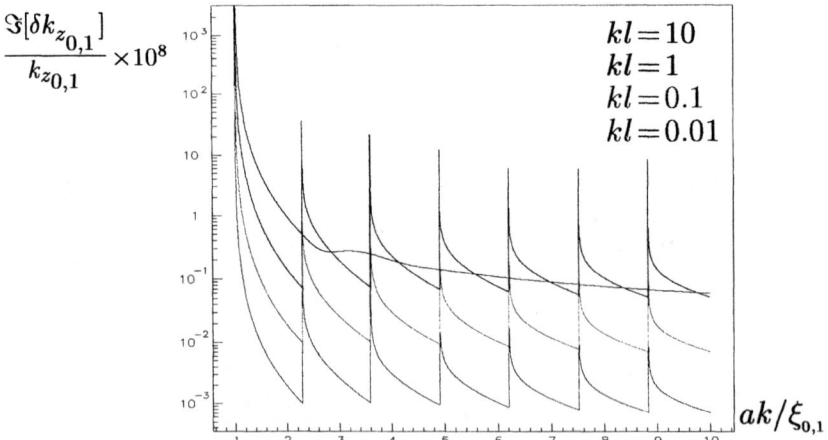

FIGURE 6. Normalized shift vs. normalized frequencies for longitudinal roughness.

CONCLUSIONS

The GFM allows us to consider multiple scattering of e.m. waves into a randomly rough waveguide, determining, as regards $TM_{n,m}$ modes of the structure, the actual shift for propagation constant δk_z for azimuthal and longitudinal corrugations. In both cases a decreasing in phase velocity emerges, depending on statistical features of random perturbation, which, within the range of validity of our approximations, does not appear sufficient to excite slow waves. For longitudinal roughness, an attenuation term has been found taking into account incoherent scattering power losses.

REFERENCES

1. K. L. F. Bane, G. V. Stupakov "Wake of a rough beam wall surface", SLAC-PUB, 8023, Dec.1998.
2. A. Mostacci, L. Palumbo, F. Ruggiero, S. Ugoli "Wake Field Effects due to Surface Roughness", VIII International Workshop on Linear Colliders, Frascati, 1999.
3. F. Bass, I. Fuks, V. Freilikher "Propagation in Statistically Irregular Waveguides. Part I: Average Field", IEEE Transactions on Antennas and Propagation, **22**-3 (1974).
4. I. M. Fuks, V. Freilikher "Green's Function Method for the Helmholtz Equation with Perturbed Boundary Conditions", Radiophysics and Quantum Electronics, **13**-1, pp.98-105 (1970).
5. F. Bass, I. Fuks "Wave Scattering from Statistically Rough Surfaces", Pergamon Press, Oxford, 1979.
6. E. Di Liberto, "Il metodo della funzione di Green per la valutazione degli effetti delle rugosità statistiche sulla propagazione di modi in guida" (in italian) , laurea thesis 1999.

A Possible Experiment at LEUTL to Characterize Surface Roughness Wakefield Effects

Sandra G. Biedron[1†], Giuseppe Dattoli[2], William M. Fawley[3],
Henry P. Freund[4], Zhirong Huang[1], John W. Lewellen[1],
Stephen V. Milton[1], Heinz-Dieter Nuhn[5]

[1]Advanced Photon Source, Argonne National Laboratory, Argonne, IL 60439, USA
[2]ENEA, Dipartmento Innovazione, C.R. Frascati, Rome, Italy 00044
[3]Lawrence Berkeley National Laboratory, Berkeley, CA 94720, USA
[4]Science Applications International Corporation, McLean, VA 22102, USA
[5]Stanford Linear Accelerator Center, Stanford University, CA 94309-0210, USA
†and Max-Laboratory, University of Lund, Lund, Sweden S-22100

Abstract. Wakefield effects due to internal vacuum chamber roughness may increase the electron beam energy spread and so have become an immediate concern for future x-ray free-electron laser (FEL) project developments such as the SLAC Linac Coherent Light Source (LCLS) and the DESY TESLA x-ray FEL. We describe a possible experiment to characterize the effects of surface roughness on an FEL driven by self-amplified spontaneous emission (SASE) operation. Although the specific system described is not completely identical to the above-proposed projects, much useful scaling information could be obtained and applied to shorter wavelength systems.

INTRODUCTION

Although the effects of internal vacuum chamber roughness have been predicted by analytical methods [1-6] and simulation [7], there is currently no experimental confirmation. Furthermore, the results are strongly dependent upon the details of the theoretical surface roughness model used. The effects may be important for future x-ray free-electron laser (FEL) projects such as the SLAC Linac Coherent Light Source (LCLS) [8] and the DESY TESLA x-ray FEL [9] that are based on the self-amplified spontaneous emission (SASE) process, as the induced energy spread could cause significant lengthening of the gain length. Currently, there are plans to test roughness effects at the SLAC Gun Test Facility (GTF) at an electron beam energy of 30 MeV [10] and between ~240-390 MeV at the TESLA Test Facility (TTF) [11-13]. In addition, we describe yet another possible experiment that could be attempted to further our understanding of surface-roughness-induced wakefields and their effect on the electron beam. To be concrete, we will consider the Low-Energy Undulator Test Line (LEUTL) at the Advanced Photon Source (APS) at Argonne National Laboratory

CP581, *Physics of, and Science with, the X-Ray Free-Electron Laser*, edited by S. Chattopadhyay et al.
© 2001 American Institute of Physics 0-7354-0022-9/01/$18.00

[14]; however, the experiment can obviously be done elsewhere. This note outlines some of the technical considerations for a roughness experiment at LEUTL.

POSSIBLE EXPERIMENTAL LAYOUT

The first LEUTL experiment examines SASE gain and saturation in three wavelength regions: the visible, ultraviolet, and vacuum ultraviolet [15-17]. This first LEUTL experiment is often referred to as the APS SASE FEL and utilizes a string of undulators that are similar to those found in the APS storage ring, each having an undulator parameter K of 3.1, a period of 3.3 cm, and a length of 2.4 m. LEUTL presently (Fall 2000) employs nine identical undulators spaced by $\cong$ 0.38 m. This space contains a combined function dipole-corrector and quadrupole magnet, electron beam diagnostics, and optical beam diagnostics [18]. LEUTL uses a photocathode-rf gun [19] and drive-laser system [20] to generate electrons that are further accelerated through the APS linear accelerator (linac) [21]. (Note: this linac is also used as part of the injector system to the APS storage ring.) The APS linac is capable of operation up to 650 MeV, corresponding to a SASE FEL output wavelength down to 59 nm. Peak currents up to 650 A have been achieved using a recently installed bunch compressor system [22], allowing a variety of experimental measurements to be obtained.

With regard to a beamline region appropriate for surface roughness wakefield effects investigations, we point out that a transfer line lies between the exit of the linac and the LEUTL tunnel proper. This line has numerous electron beam diagnostics, the necessary corrector and quadrupole magnets to insure a proper transport to and match into the LEUTL tunnel, and a high-resolution electron beam spectrometer capable of resolving one part in a thousand. There is sufficient space upstream of this spectrometer to install various vacuum chambers with a variety of induced surface roughnesses. These would be mounted on an actuator allowing the insertion or removal of the chamber under study. Furthermore, there will be an identical spectrometer installed at the exit of the linac within a year, thus allowing one to measure the electron beam energy and energy spread both before and after traversing the chamber in question. The existence of two spectrometers is the essence of this proposal in that it would allow a clean subtraction of all other wakefield effects that might arise in the transport between the two identical spectrometers. Also, there is a high-resolution energy spectrometer after the exit of the undulator line, so the system can also test the effect of the wakefields from the test pipe on the SASE output. The general layout is shown in Figure 1.

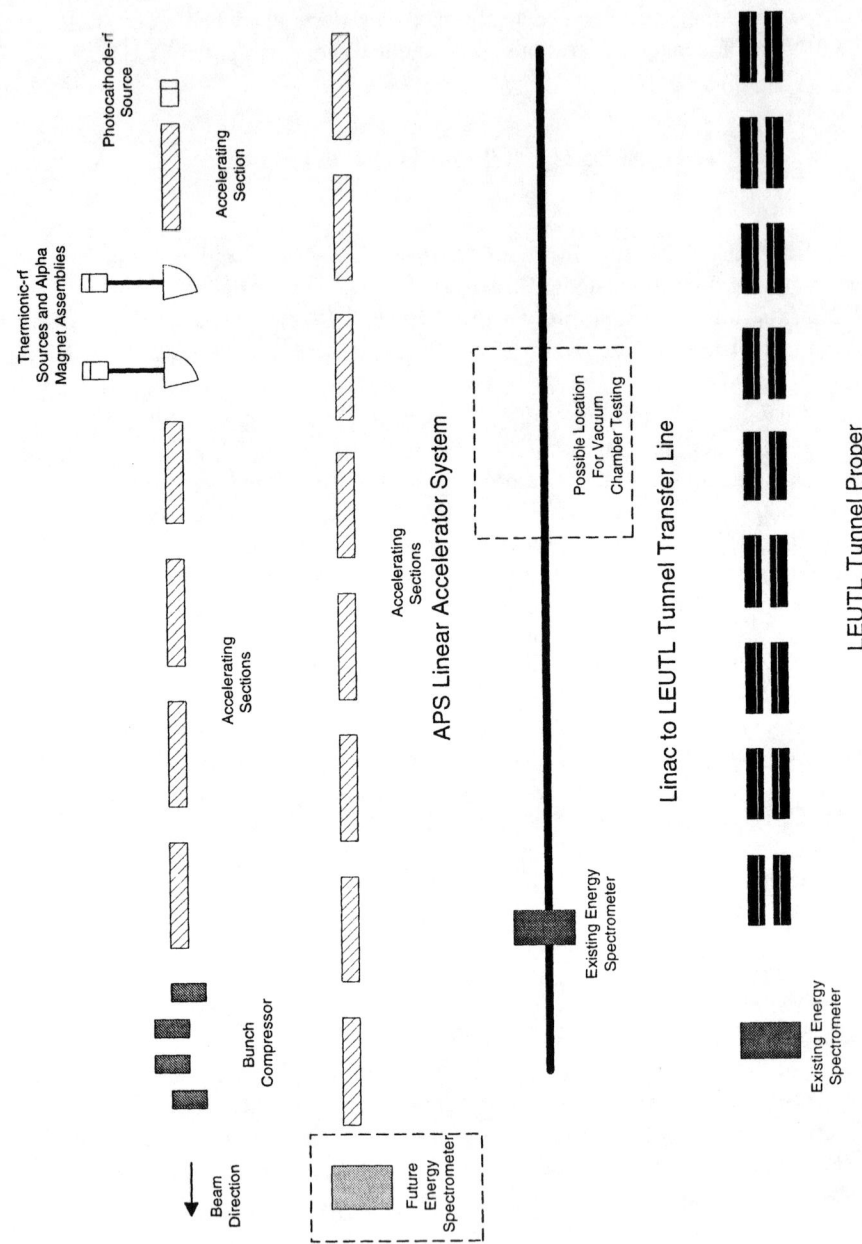

FIGURE 1. LEUTL layout with notation of place for future spectrometer and possible place for test chamber installation.

WHAT ROUGHNESS TO IMPOSE?

A number of different roughness models have been developed and published by authors such as K. Bane, G. Stupakov, A. Novokhatski, A. V. Agafonov, L. Palumbo, and others [1-6]. The effects depend upon the bunch length, the radius, and the length of the vacuum pipe as well as details of the surface roughness.

Surface roughness experiments are not likely to realize either the bunch length (about 100 fs or less) or the length of the vacuum pipes (100 m or more) that are proposed for the LCLS and the TESLA FEL. With longer bunches and shorter vacuum chambers, the effects from realistic surfaces would be too small to be measured in a test experiment. It will therefore be necessary to prepare a special beam pipe with artificially produced large-scale roughness structures on the inside.

The roughness models, discussed in the literature, are based on two types of geometrical structures: small bumps with heights that are equal to the distance between them or grooves that are longitudinally elongated with respect to their depth. The structures that can be purposely created on the inside surface of a long, narrow vacuum tube (typically 5 mm to 1 cm in diameter) are limited. The GTF group decided, therefore, to restrict their test to a surface structure with a periodicity on the order of the bunch length and transverse groove lengths of about half that. Theory predicts that for such an arrangement the induced energy spread scales proportionally to the length of the pipe and inversely to the square of the pipe radius.

An experiment at LEUTL could use a similar strategy but would have access to a wider range of bunch lengths due to the existence of the bunch compressor. A scanning electron probe, with a resolution on the atomic scale horizontally and 5 Å vertically, is available at the Advanced Photon Source and could be used to accurately measure the surface roughness of the sample pipes. Also, to achieve a more capable method of producing the artificial roughness as well as to properly measure it, there is some thought of splitting the chamber in two 180 degrees apart with fitting grooves. This entire assembly would be placed on an actuator inside a larger vacuum chamber. Different chambers could then be inserted and removed frequently without disruption of the vacuum system. As stated earlier, a direct measurement of SASE degradation as a function of different test chambers could be made at LEUTL, as it is presently configured as an operating SASE FEL.

CONCLUSION

An experiment could be constructed to clearly show the effect of surface roughness on an electron beam. The LEUTL at the APS was used as an example. Since this experiment relies upon the successful installation of a high-resolution electron beam spectrometer, the earliest this experiment could be scheduled at the LEUTL facility is the summer of 2001. Experimental comparisons of the theoretical models and a scaled prototype chamber would determine if the predictions are strongly model dependent and help in the design of upcoming x-ray free-electron laser systems. At present, we are awaiting input from our theoretical colleagues. With their assistance, proper scaling of the effects to our lower energy, longer bunch length systems could be done to determine the types of vacuum chambers we should be testing. This information could then be used to directly predict the effect on projects such as the LCLS and TTF x-ray facilities.

ACKNOWLEDGMENTS

The work of S. G. Biedron, Z. Huang, J. W. Lewellen, and S. V. Milton was supported at Argonne National Laboratory by the U.S. Department of Energy, Office of Basic Energy Sciences under Contract No. W-31-109-ENG-38. The activity and computational work for H. P. Freund was supported by Science Applications International Corporation's Advanced Technology Group under IR&D subproject 01-0060-73-0890-000. The activity of W. Fawley was supported at Lawrence Berkeley National Laboratory by the U.S. Department of Energy under Contract No. DE-AC03-76SF00098. The work of H.-D. Nuhn was supported by the U.S. Department of Energy, Office of Basic Energy Sciences, Division of Material Sciences, under Contract No. DE-AC03-76SF00515.

REFERENCES

1. Bane, K., Ng, C., and Chao, A., "Estimate of the impedance due to wall surface roughness," SLAC-PUB-7514, 1997.
2. Stupakov, G., "Impedance of small obstacles and rough surfaces," PRST-AB, I, 1998, 064401.
3. Stupakov, G., Thomson, R. E., Walz, D., and Carr, R., "Effects of beam-tube roughness on X-ray free electron laser performance," *Phys. Rev. ST Accel. Beams* **2**, 1999, 060701.
4. Novokhatski, A., Timm, M., and Weiland, T., "The surface roughness wakefield effect," Proceedings of the ICAP '98, Monterey, California, 1998.
5. The A.V. Agafonov (P.N. Lebedev Physics Institute, Moscow) model, as discussed in one of this workshop's sessions and summaries.
6. Palumbo, L., Angelici, M., Frezza, F., Mostacci, A., and Spataro, B., "Wake Fields Due to Periodic Roughness in a Circular Pipe," Proceedings of EPAC 2000, Vienna, Austria, 2000, pp. 1438-1440.
7. Reiche, S., and Schlarb, H., "Simulation of time-dependent energy modulation by wake fields and its impact on gain in the VUV free electron laser of the TESLA Test Facility," *Nucl. Instrum Meth A* **445**, 2000, pp. 155-159.
8. The Linac Coherent Light Source (LCLS) Design Study Group, "LCLS – Design Study Report," Stanford Linear Accelerator Center publication number SLAC-R-521, April 1998.
9. "A VUV Free Electron Laser at the Tesla Test Facility at DESY. Conceptual Design Report," DESY Print, TESLA-FEL 95-03, June 1995.
10. Nuhn, H-D., private communication.
11. Dohlus, M., Lorentz, R., Kemps, Th., Schlarb, H., and Wanzenberg, R., "Estimation of the longitudinal wakefield effects in the TESLA-TTF FEL undulator beam pipe and diagnostic section," TESLA-FEL 98-02, March 1998.
12. Novokhatski, A., Timm, M., and Weiland, T., "A Proposal for the Surface Roughness Wake Field Measurement at the TESLA Test Facility," Proceedings of the IEEE 1999 Particle Accelerator Conference, 1999, pp. 2879-2881.

13. Hüning, M., Schlarb, H., and Timm, M., "Experimental setup to measure the wake fields excited by a rough surface," *Nucl. Instrum. Meth. A* **445**, 2000, pp. 362-365.
14. Milton, S. V., "The low energy undulator test line," in Free Electron Laser Challenges, Patrick G. O'Shea and Harold E. Bennett, Editors, Proceedings of SPIE, **2988**, 1997, pp. 20-27.
15. Milton, S. V., et al., "Status of the Advanced Photon Source low energy undulator test line," *Nucl. Instrum. Meth A* **407**, 1998, pp, 210-214.
16. Milton, S. V., et al., "The FEL Development at the Advanced Photon Source," in Free Electron Laser Challenges II, Harold E. Bennett, David H. Dowell, Editors, Proceedings of SPIE, **3614**, 1999, pp. 96-108.
17. Milton, S. V., et al., "Observations of Self-Amplified Spontaneous Emissions and Exponential Growth at 530nm," *Phys. Rev. Lett.* **85,** 2000, pp. 988- .
18. Gluskin, E., et al., "The magnetic and diagnostics systems for the Advanced Photon Source self-amplified spontaneously emitting FEL," *Nucl. Instrum. Meth. A* **429**, 1999, pp. 358-364.
19. Biedron, S. G., et al., "The Operation of the BNL GUN-IV Photocathode Gun at the Advanced Photon Source," Proceedings of the IEEE 1999 Particle Accelerator Conference, 1999, pp. 2024-2026 and references therein.
20. Travish, G., Arnold, N., and Koldenhoven, R., "The Drive Laser for the LEUTL FEL RF Photoinjector," Proceedings of the Twenty-first International Free Electron Laser Conference August 23-26, 1999, Hamburg Germany, 2000, pp. 101-102.
21. White, M., et al. "Construction, Commissioning and Operational Experience of the Advanced Photon Source (APS) Linear Accelerator," Proceedings of the XVIII International Linear Accelerator Conference, Geneva, Switzerland, 1996, pp. 315-319.
22. Borland, M., Lewellen, J., and Milton, S., "Highly Flexible Bunch Compressor for the APS LEUTL FEL," Proceedings of the XX International Linear Accelerator Conference, Monterey, California, August 21-25, 2000 (to be published).

A Method for Increased Resolution of X-ray Images

Sergio Monteiro

Department of Physics
Moorpark College
7075 Campus Road
Moorpark, CA 93021
USA
monteiro@physics.ucla.edu

Abstract. Traditional X-ray imaging systems mixes transmitted with scattered radiation. The scattered radiation in the traditional method only degrades image quality. These systems may add selective absorbers of scattered radiation to improve image quality (Bucky grids). We describe a scanning beam method that generates good images with both the transmitted as well as the scattered radiation. The method described here generates two separate images, either one of better resolution than the traditional method. We include mathematical modeling examples of both the traditional as well as the proposed system.

TRADITIONAL X-RAY IMAGING SYSTEMS

Standard X-ray Systems

Most of us have taken several X-ray pictures and are more or less familiar with the system. Essentially similar systems are used for industrial non-destructive testing. A broad X-ray beam is produced towards the object to be imaged, a person, for example. The object must be composed of parts with dissimilar X-ray absorption and scattering cross-sections. The mental picture most people make of the image so created includes only the transmitted radiation: the higher density parts of the object (bones, for example) absorb more, and consequently the "shadow" picture recorded on a film behind the object receives less photons behind these higher density areas. We all know about scattering, and will include it in the scheme if asked to do so, but we correctly implicitly assume that scattering is a minor factor, so we justifiably leave this part aside. The majority of X-ray pictures require low resolution and consequently the neglect of scattering is acceptable. For higher resolution pictures, selective absorbers of scattered radiation are inserted between the object and the recording film, with the objective of partial elimination of the scattered radiation. These selective absorbers are known as Bucky grids.

CP581, *Physics of, and Science with, the X-Ray Free-Electron Laser,* edited by S. Chattopadhyay et al.

Analysis of the Quality Degradation of Standard X-ray Imaging Systems

Standard X-ray systems produce a density image on a photographic film, which is placed behind the object that is irradiated with a broad, approximately collimated, X-ray beam. The portions of the film that are directly behind higher density portions of the object receive less radiation than other areas of the film that are directly behind smaller density portions of the object. This is what we will call the transmitted image for simplicity; it is a shadow image of the object density. It is also a negative shadow image. Indeed, contrariwise to an ordinary shadow, which is darker behind the object, the transmitted image is more transparent behind the object, or towards the white. The X-ray transmitted image is such that higher density parts are imaged more transparent, or towards the white.

Superimposed on the same transmitted image (produced by radiation transmitted through the object) is the scattered radiation which also produces a sort of an image. The scattered radiation is necessarily at points on the recording film that are around – not behind - the projected shade of the feature that is responsible for the scattering. The scattered radiation is a positive record of the object density, as the larger the density the larger is the intensity of the scattered radiation and the darker or more opaque is the impression on the film. The scattered radiation necessarily degrades the image created with the transmitted radiation both because it is a positive image that is at odds with the main, transmitted image, also because it is spread around the feature. Accordingly, all attempts to deal with the scattered radiation have been directed to eliminating it. But this needs not be so, as it will be shown in the sequel.

Superposition of the scattered radiation is not the sole cause of image degradation in standard systems. Standard X-ray systems also implicitly make an averaging, or smearing of both images: transmitted and scattered. This averaging is a consequence of two factors: film granularity and the finite size of the source. The latter factor is the worst offender. Each point on the object is illuminated by photons that originated from different points on the X-ray source. Each of these points on the X-ray source originates beams that cross at the object from slightly different angles then proceed to different points on the photographic film. This way the same point on the object is imaged over a small region around the projection of the point on the photographic film. It is an effect of the source finite size.

NEED FOR HIGHER RESOLUTION X-RAY IMAGING

There are some classes of problems that require, or could make excellent use if available, substantially higher resolution. One such area is mammography. It is known that micro calcifications are precursors of female breast cancer. These micro calcifications are currently detected only after they reach a size of the order of 50 micrometers. This 50 micrometers size was not determined to be the optimum calcification size to indicate possible problems in the future; it simply is the smaller calcification size that can be detected with current X-ray systems. A system that were able to resolve smaller calcifications would offer the possibility of earlier detection, which could, in principle, increase the survival rate, as experience has shown that early intervention improves the likelihood of successful treatment of female breast cancer.

Another area that would profit from higher resolution images is non-destructive testing. In this case a system produces an X-ray image of a manufactured piece, an aircraft turbine blade, a fuselage part, or a critical nuclear reactor part for example. In some cases the X-ray image may reveal internal imperfections: cracks, bubbles, etc. that could compromise the part integrity.

DESCRIPTION OF THE NEW SYSTEM

Images with better resolution than the images currently available can be made with a scanning beam. Referring to Figure 1, the X-ray beam labeled 230 scans the object 20 while a pixel array detector 240 moves behind the object, following the scanning beam. The scanning process consists in pointing the X-ray beam on a particular direction, stop the beam, measuring and recording the values from the detector, followed by a (small) displacement of the beam and the detector behind it, repeating this process over the whole cross section of the object. As the X-ray beam 230 scans object 20, the central pixel on detector 240, which follows the scanning beam, continuously measure the intensity of the transmitted radiation, while the peripheral detectors measure the intensity of the scattered radiation. As the beam scans the object 20 under the control of a computer 250, image points are acquired by the pixel detector 240 that follows the motion of the X-ray beam.

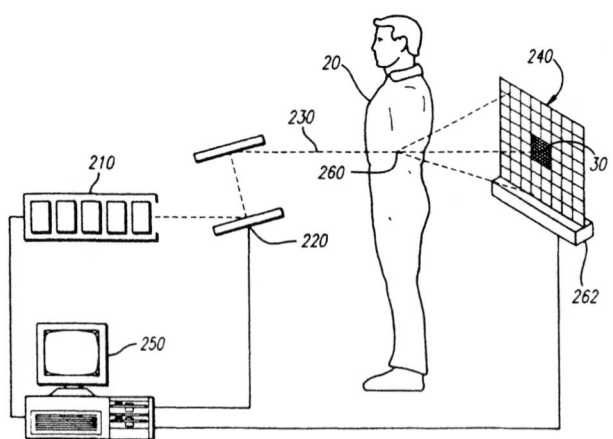

FIGURE 1. The three parts of the scanning beam system (a wiggler X-ray source 210, steering mirrors 220 and the detector 240) are controlled by a computer.

The detector 240 records both the transmitted radiation at the central pixel and the scattered radiation on the peripheral pixel detectors as well. The readings from the former (central pixel readings) are accumulated on a 2-dimensional matrix which contains the object's X-ray density, while the readings from the latter (peripheral pixels) are accumulated on another 2-dimensional matrix, that contains also the object's X-ray density, but due to scattering, as opposed to transmission. The former is the transmitted matrix T_{ij}, the latter is the scattered matrix S_{ij}. The transmitted matrix T_{ij} defines the transmitted picture, the scattered matrix S_{ij} defines the scattered image.

Notice that when the beam and central pixel detector are momentarily along the direction of point P_{ij} (x_i, y_j), an entry is made for T_{ij} (ij entry on the transmitted matrix) which is proportional to the reading of the central pixel detector. The corresponding entry on the scattered matrix Sij is the integrated readings of all the peripheral pixels around (x_i, y_j), when the beam (and central pixel detector) are at position $P_{ij} = (x_i, y_j)$. Saying it from another point of view, the entries of the scattering matrix S_{ij} are the integrated readings not of the pixel detectors at (x_i, y_j), but actually of the detectors that are *not* at (x_i, y_j). The scattered matrix is related to the beam position, *not* to the position of the pixels that originate S_{ij}. Consequently each entry (i,j) on both the transmitted matrix T and on the scattered matrix S are both related to the *same* line along the object 20 and are related to each other.

Naturally that T_{ij} differs from S_{ij} in that while a plot of the former produces a negative picture of object 20 X-ray density (atomic number), a plot of the latter produces a positive picture of object 20 X-ray density. Gray scale plots of the entries of matrices T and S are to each other as ordinary 35 mm negatives and prints made from them are to each other (see Figures 2 a and b). Addition of the entries of matrices T and S to produce a combined matrix T+S does not produce a sharper gray scale image but actually a fuzzier one. On the other hand, a gray scale plot of the matrix NT (negative of T) = 1 – T produces a picture that is in every respect similar to a gray scale plot of matrix S. 1 – T is the complement of T, which corresponds to the print that we keep in the album, not to the negative that we store away.

It follows that matrices T and S are both genuine sharp images of the X-ray density of the object 20. For the sake of familiarity, radiologists that are used to analyze negative images will prefer to look at T and NS = 1 – S, the negative of S, both of which have the same negative appearance. Naturally that a single picture can be created with T + NS (which is a negative picture) or with S + NT (which is a positive print so to say).

Neither of these individual pictures (T and S) are impacted by the destructive interference of the images generated by current systems, nor is the negative image T + NS or the positive (print) S + NT produced with the combined information.

Two matrices are created, the former with what we call transmitted radiation matrix, the latter with what we call scattered radiation matrix. If the former matrix (transmitted radiation matrix) is plotted on a gray scale proportional to the numerical values of its entries, it will produce a negative image of the object's X-ray opacity, much like an ordinary X-ray picture, but without any scattering information superimposed on it. If the latter matrix (scattered radiation matrix) is plotted on a gray scale proportional to the numerical values of its entries, it will produce a positive image of the object's X-ray scattering cross-section. One can also visualize the transmitted image as an ordinary 35 mm negative produced with visible light, and the scattered image as an ordinary print produced from the same negative.

MATHEMATICAL MODELING

Figures 2 (2a and 2b) and 3 (3a and 3b) are mathematical modeling of the images from an artificial object with three bone-like features: a 1 micrometer diameter calcification, a 2 micrometer diameter calcification and a vertically elongated 11 micrometers wide calcification. Figures 2a and 2b display a modeling of the scanning beam method for the worse situation of maximum spatial quantization error.

Figures 3a and 3b display modeling of a modern but standard system with conservative smearing of 4 and 8 micrometers which are smaller than what is possible to achieve in reality with existing systems. No standard X-ray systems is as good as the modeling at Figure 3. This smearing is a combination of film granularity and X-ray source finite size. It follows that these modelings compare a worse case for a scanning beam method with a best case for a modern standard system. Notice that the scales on these simulated pictures are beyond the resolution of standard systems, the whole area being 100 micrometers long by 70 micrometers wide, corresponding to the width of one-and-a-half hair filaments, the calcifications themselves being invisible to the naked eye. The whole area on Figures 2 and 3 is visible to the naked eye but are very small, while the two small calcifications at 20 and 40 micrometers are only visible with a microscope.

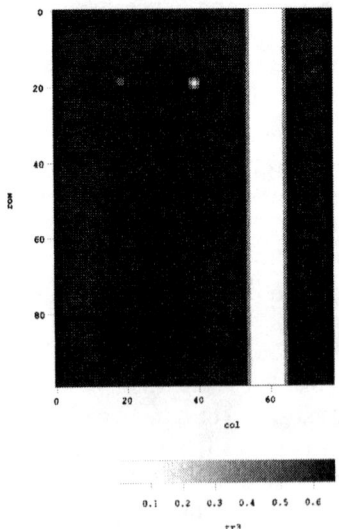 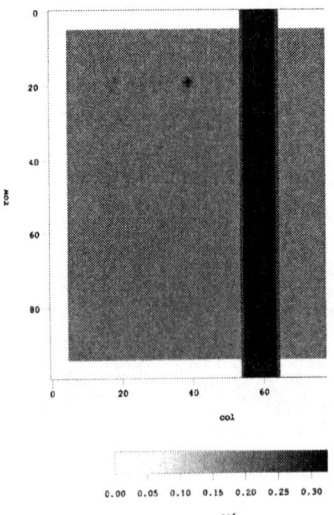

FIGURE 2. Simulated images from the scanning beam method. This is a worse case simulation of a 1x1 micrometer calcification at x=20 micrometers, a 2x2 micrometer calcification at x=40 micrometers, and an 11 micrometers wide long calcification at x=60 micrometers. (a) Transmission, (b) Scattering.

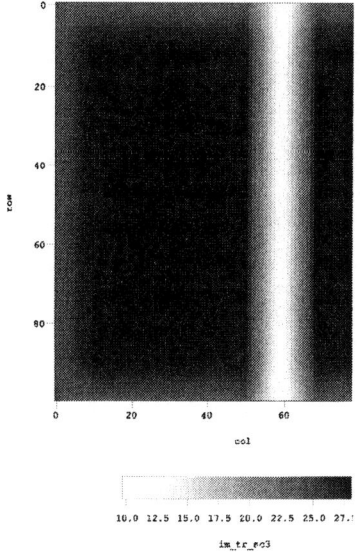

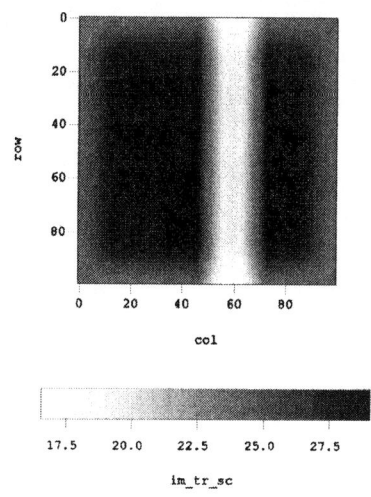

im_tr_es3

im_tr_sc

FIGURE 3. Simulated images from standard existing systems. The objects are the same as in Figure 2. 3a assumes a smearing of 4 micrometers while 3b assumes a smearing of 8 micrometers, both cases less than what can be achieved at present time.

The values used in the computer simulation for film density on an X-ray film are from measurements by the author on authentic medical X-ray films. Relationships among the different parameters can be found, e.g., in [1]. For sake of simplicity all the modeling was calculated on x-y without corrections for the radial characteristics. Nor did we include angular dependence of scattering – all scattering is calculated as having the same intensity on all directions and all angles. The modeling allow scattering at arbitrary angles but without intensity dependence on the scattering angle – a simplification. This simplification does not change the fundamental nature of the modeling, only details. Specific details for the construction of a scanning beam system can be found in [2].

CONCLUSION

This new method to create X-ray images with a scanning beam allows better resolution than the current systems. The new method requires a thin, steerable beam, as provided by an X-ray laser for example. The resolution improvement of the scanning beam method over existing systems is better than one order of magnitude.

ACKNOWLEDGMENTS

I am indebted to financial support from Moorpark College, particularly Vice-Presidents Eva Conrad and Ruth Hemming and the Dean of Sciences, Floyd Martin, and to the Academic Affairs Committee, particularly to Lori Bennet.

REFERENCES

5. Bushong, S.C., *Radiologic Science for Technologists*, fourth edition, C.V.Mosby Company (1988).
6. Monteiro, S., *Radiation Imaging System*, US Patent # 5,590,169, 31 December 1996

PREDICTED PERFORMANCE OF THE LCLS X-RAY DIAGNOSTICS

E.Gluskin, P.Ilinski and N.Vinokurov

Advanced Photon Source, Argonne National Laboratory, 9700 S. Cass Ave., Argonne, IL 60439, USA

Abstract. An x-ray diagnostics concept is proposed for the Linac Coherent Light Source (LCLS),. X-ray diagnostics will provide spectral and spatial measurements of the spontaneous and SASE radiation along the undulator line, and will be complimentary to an electron beam diagnostics. Main goal of the x-ray diagnostics is to measure the radiation gain along the undulator line. Diagnostics setup consists of diamond crystal monochromator and silicon PIN diode and CCD camera detectors. Accuracy and dynamic ranges of x-ray diagnostics are discussed.

INTRODUCTION

The purpose of the LCLS x-ray diagnostics is to support and verify independently from the electron beam-based alignment procedure the performance of the LCLS. The x-ray diagnostics consists of tools for measurements and analysis of the spectral and spatial characteristics of spontaneous and SASE radiation along the undulator line. The main goal is to measure the absolute flux of x-rays as a function of the distance along the undulator line. Furthermore, the diagnostics permits to verify overlapping cones of undulator radiation from different segments.

Two cases of the x-ray diagnostics are considered – on-axis and off-axis. On-axis diagnostics is performed at the photon energies close to the fundamental energy of undulator radiation. In this case electron beam will hit the monochromator's crystal. In the other case an off-axis, "red-shifted" part of undulator radiation is utilized.

Structure of the undulator cell

The LCLS undulator line is designed as a set of standard cells shown schematically in Fig.1. Each cell consists of three 3.24-m long undulator segments with two 187 mm separation breaks and one 421 mm separation between undulators. The short breaks are filled with the focusing or defocusing lens, electron beam position monitors (BPM) and steering coil. The last break includes additional x-ray diagnostics set-up.

The length of the undulator has been optimized based on the requirement to operate the LCLS at both 4.5 GeV and 14.5 GeV electron beam energies. Also, it has been shown independently that chosen cell length corresponds to the optimum position of BPMs [1].

X-ray diagnostics specifications and experimental set-up

In order to be efficient and meaningful x-ray diagnostics should satisfy following requirements:
- all types of spectral, flux and spatial measurements should be sensitive enough to get accurate information from one shot;
- spatial measurements should provide angular resolution of several microradians;
- accuracy of the absolute flux measurements should be better than 10%;
- diagnostics should have wide dynamic range of many orders of magnitude.

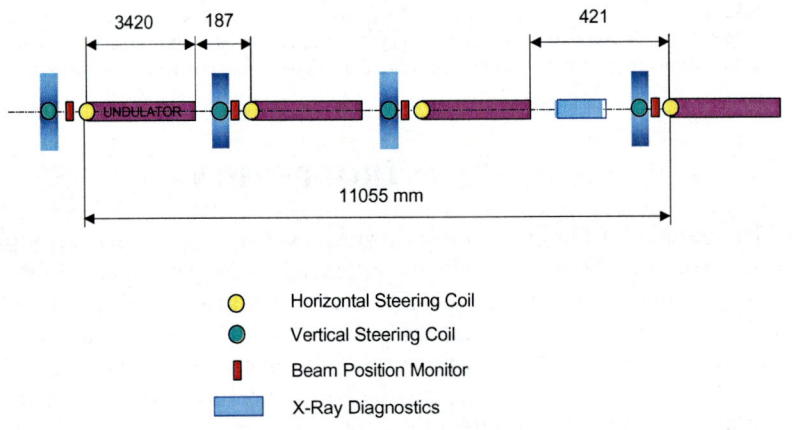

FIGURE 1. Structure of a standard cell.

The experimental set-up that meets the above requirements is shown in Fig.2. It consists of 200-μm-thick diamond (111) crystal monochromator, x-ray CCD cameras and the PIN diode.* The monochromator has ~10^{-4} energy bandwidth and can perform

---

* The diagnostics set up for the LCLS undulator line in principle is quite similar to one for the Low Energy Undulator Test Line (LEUTL) at the APS [2]. Although LEUTL diagnostics was developed for the visible light whereas LCLS diagnostics aims x-ray.

in the energy range between 4 to 9 keV. Combination of the monochromator and the CCD camera as an area detector are used for spatial (angular) measurements, whereas the combination of the monochromator with the PIN diode provides the absolute flux measurements along the undulator line.

ON-AXIS X-RAY DIAGNOSTICS

The purpose of the on-axis x-ray diagnostics is to measure in absolute units the increase in the spectral flux along the undulator line and to provide information about the spatial distribution of radiation. In order to analyze performance of the on-axis x-ray diagnostics a set of calculations of undulator radiation from the cell has been conducted using the program SRW [3].

For on-axis x-ray diagnostics, the monochromator is set near the fundamental undulator harmonic at the energy of 8.27 keV, which corresponds to a Bragg angle of 21.35 degree. Calculations of the spectral flux generated by the undulator cell and transmitted through the monochromator yield $5 \cdot 10^6$ photons/shot, charge of 1.6 nC will be registered by the silicon PIN diode. A cooled silicon PIN diode could detect a single x-ray photon. Using combination of PIN diode registration setup, the dynamic range of ten orders of magnitude could be covered and no additional attenuation of x-ray beam will be required.

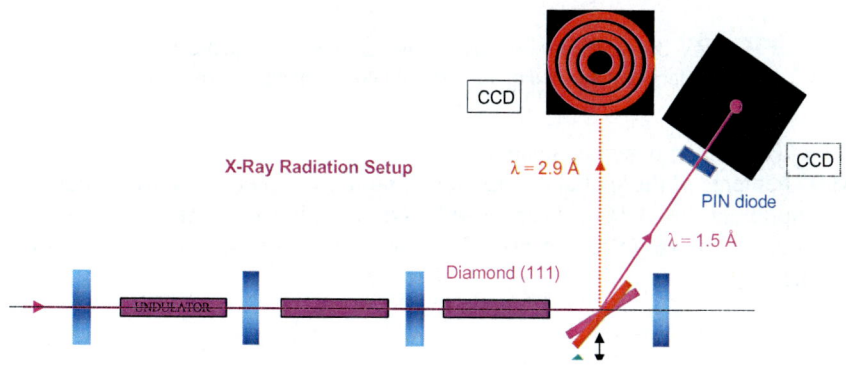

FIGURE 2. Diagnostics Setup.

X-ray diagnostics setup will provide absolute (within 10%) measurements of spectral flux after each undulator cell. The growth rate could be derived from the measured flux at consequent diagnostics stations. The growth rate of the flux is a reliable source for evaluation and study of the development of the SASE process.

The spatial flux distribution for one undulator cell is the superposition radiation from three undulators. In order to evaluate the sensitivity of the x-ray diagnostics to the angular misalignment of undulators, e-beam trajectories in the first and third undulators in the cell have by a missteering angle θ_{mis} (Fig.3). A series of calculations of spontaneous radiation for nominal electron beam emittance of 0.05 nm·rad at photon energy of 8.29 keV (detuned slightly from the 8.27 keV fundamental energy to higher energy) and at a distance of 60 m from the undulator cell have been performed. Calculations performed at the detuned from the fundamental energy because in this case angular distribution of the undulator radiation is smaller compare to that at the fundamental energy. Moreover, by detuning to the higher energies from the fundamental will allow to decrease intensity in order to accommodate the dynamical range of the detectors. The results of these calculations are shown in Figs. 4 and 5. It is quite obvious from these data that at least 4 µrad angular resolution could be easily achieved if one would use the diagnostics station far enough from the undulator cell. During alignment procedure trajectory bumps could be introduce at particular undulator

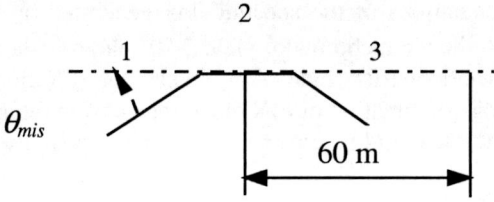

FIGURE 3. Electron beam trajectory through the three undulators in the cell, trajectory is missteered in the first and third undulators by angle θ_{mis}.

in order to clarify an observed image.

Measurements of the spatial distribution of radiation generated in each undulator cell will compliment the electron beam-based alignment, but will not substitute it. X-ray diagnostics will be very useful especially in the first steps of the beam-based alignment procedure.

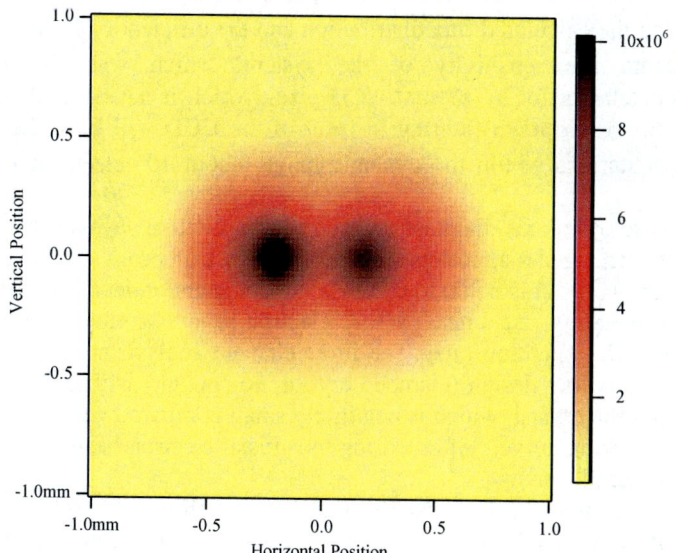

FIGURE 4. Calculated spatial distribution of the undulator radiation from three-undulator cell with a missteering θ_{mis} = 4 μrad, electron beam emittance of 0.05 nm·rad, and photon energy of 8.29 keV, at 60 m from the undulators.

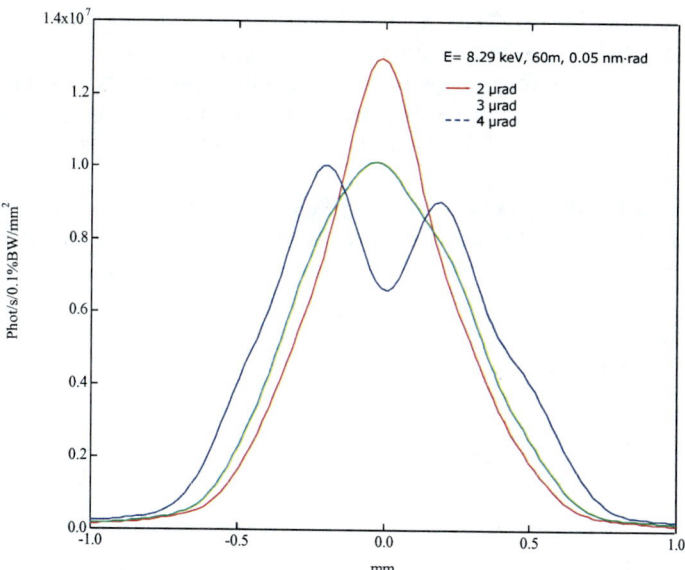

FIGURE 5. Calculated horizontal profiles of the undulartor radiation for missteering angles θ_{mis} of 2, 3 and 4 μrad, electron beam emittance of 0.05 nm·rad, and at photon energy of 8.29 keV, at 60 m from the undulators.

Knowing the calculated flux distribution and the efficiency of x-ray diagnostics one can estimate the sensitivity of the system, which was found to be $5 \cdot 10^2$ electrons/pixel/shot for a 7x7 μm^2 CCD pixel, which is substantially above the CCD noise level. Appropriate filtering in front of the CCD will keep the flux in the last diagnostics stations within the dynamic range (about 10^5 electrons per pixel) of the detector.

In the case of on-axis diagnostics electron beam will always hit the diamond crystal in one of the engaged diagnostics stations. The electron beam energy loss in a 200-μm-thick diamond crystal is equal to 0.25 MeV/particle and independent of the 4.5 or 14.5 GeV particle energy. As a result a 1 nC electron beam introduces average power of 30 mW for 120 Hz repetition rate. The finite element analysis shows that the use of the most simple cooling design (clamped crystal, no coolant) will lead to a slope error of 0.06 μrad on the crystal, which is negligibly small compared with the 35 μrad width of the crystal rocking curve. After exiting the crystal electron beam will have an angular spread of 40 μrad (rms).

Recent experimental data taken at the FFTB facility at SLAC show that diamond crystal withstands high energy, highly focused electron beam without any visually observed damage [4].

OFF-AXIS X-RAY DIAGNOSTICS

The off-axis, "red-shifted" x-ray diagnostics is complimentary to the on-axis diagnostics. The crystal monochromator in this case may have a small hole to let the electron beam go through unperturbed. It allows observing radiation from each undulator cell without any trajectory distortion. A similar technique was successfully implemented for the LEUTL diagnostics, where a mirror was used instead of the crystal.

For the "red-shifted" radiation of 4.25 keV, the crystal monochromator has a Bragg angle of 45 degree and the angle between the CCD camera and radiation is 90 degree. The results of the calculated spatial distribution of undulator radiation from two undulators at distance of 6 m and with missteering angle θ_{mis} of 10 μrad are shown in Figs. 6 and 7. In general, the angular resolution of the off-axis diagnostics is lower compare to the on-axis one.

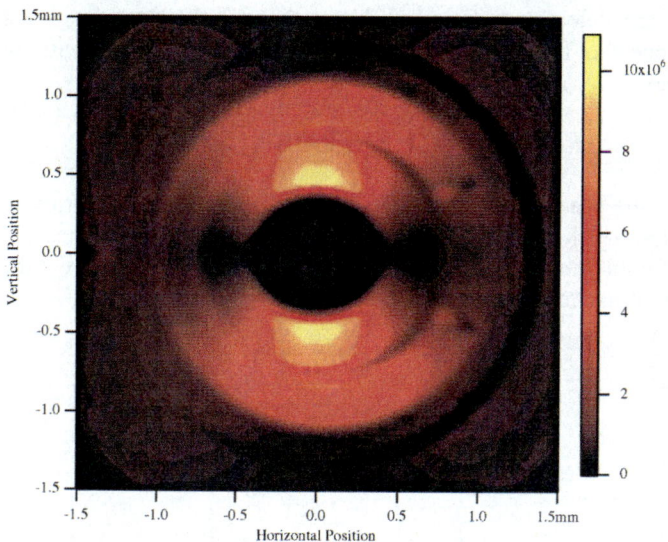

FIGURE 6. Calculated spatial distribution at energy of 4.25 keV from two undulators at distance of 6 m, and with missteering angle θ_{mis} of 10 μrad between undulators.

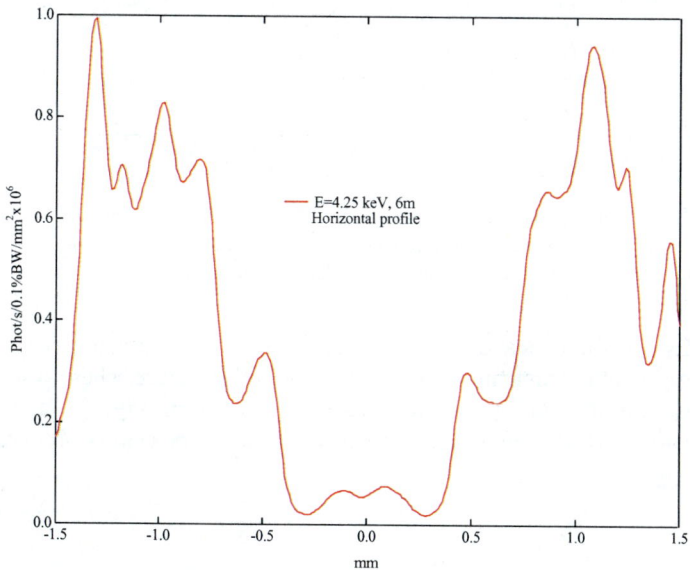

FIGURE 7. Calculated horizontal profile of spatial distribution at energy of 4.25 keV from two undulators, at 6m from undulators, and with missteering angle θ_{mis} of 10 μrad between undulators.

Calculation of the detection efficiency shows that CCD camera will get $4 \cdot 10^2$ electron/pixel per one shot for the 20 x 20 μm^2 pixel, which is again substantially above the CCD noise level.

EMITTANCE INFLUENCE

The influence of the beam emittance on the spatial distribution of radiation has been studied. The results are shown in Figs. 8, 9 and 10. Horizontal profiles of spatial distribution for different emittances at energy of 8.29 keV for the cell structure where first and third undulators are misaligned by 4 μrad each is shown in Fig. 8.

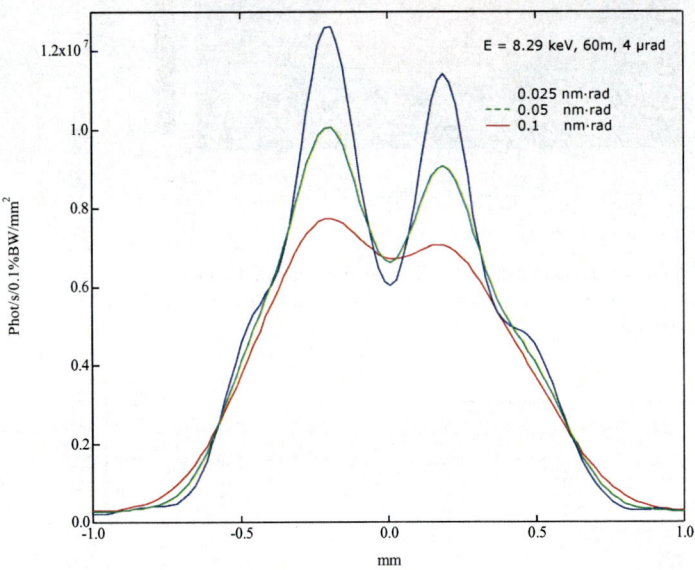

FIGURE 8. Calculated horizontal profiles of undulator radiation from two undulators at energy of 8.29 keV with missteering angle θ_{mis} of 4 μrad, at 60 m from undulators and for electron beam emittances of 0.025, 0.05 and 0.1 nm·rad.

A spatial intensity distributions and horizontal profile at 60 m and photon energy of 8.3 keV (high energy detune from fundamental) for the cell structure where first and third undulators are misaligned by 10 μrad each is shown in Fig. 10 and 11. Differences between 0.025, 0.05 and 0.1 nm·rad emittances could be clearly observed.

138

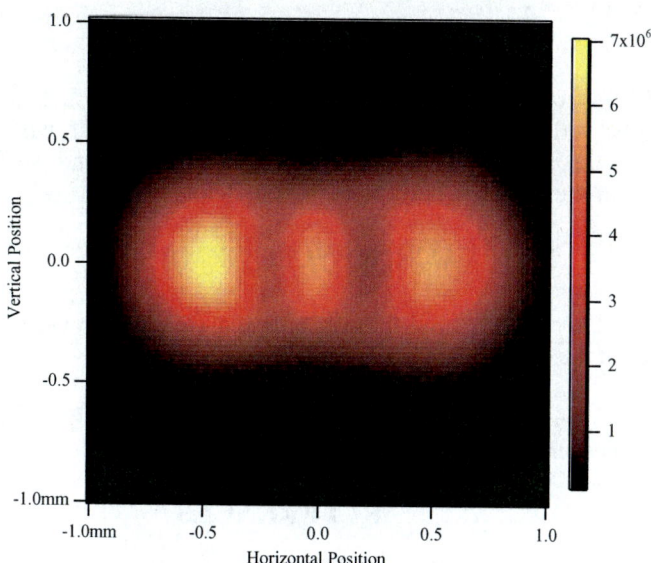

FIGURE 9. Calculated spatial distribution of the undulator radiation from three-undulator cell with a missteering angle θ_{mis} of 10μrad, at energy of 8.3 keV (slightly detuned from the 8.27 keV fundamental to higher energy), at 60 m from undulators and with electron beam emittances of 0.05 nm·rad.

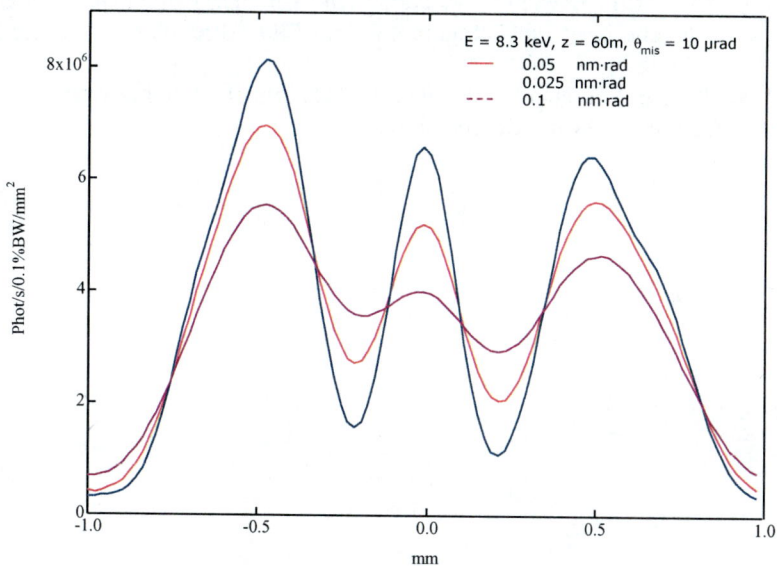

FIGURE 10. Calculated horizontal profiles of the undulator radiation from three-undulator cell with missteering angle θ_{mis} of 10μrad, 8.3 keV (slightly detuned from the 8.27 keV fundamental to higher energy), at 60 m from undulators and electron beam emittances of 0.025, 0.05 and 0.1 nm·rad.

CONCLUSIONS

The LCLS x-ray diagnostics will be a valuable tool that will support and independently verify the electron-beam-based alignment. It will also provide a feedback for the fine-tuning of the undulator matching and verification of the development of the SASE process along the undulator line.

ACKNOWLEDGEMENTS

We thank Dr. L.Emery and G.Tajiri for the set of calculations related to the performance of the diamond monochromator.

This work is partially supported by the U.S. Department of Energy, Office of Basis Energy Sciences, under Contract No. 31-109-Eng-38.

REFERENCES

1. P. Emma, Electron Phase Slip in an Undulator with Dipole Field and BPM Errors, These Proceedings

2. E. Gluskin, C. Benson, R.J. Dejus, et al., The magnetic and diagnostics systems for the APS SASE FEL, NIM, A 429 (1999), pp.358-364

3. O. Chubar, P. Elleaume, "Accurate And Efficient Computation Of Synchrotron Radiation In The Near Field Region", Proc. of the EPAC98 Conference, 22-26 June 1998, p.1177-1179.

4. W.-K. Lee, R. Iverson, P. Krejcik, D. McCormick, Summary on the diamond test with SLAC focused beam, October 2000

Surface Roughness Impedance

G. V. Stupakov

Stanford Linear Accelerator Center Stanford University, Stanford, CA 94309

Abstract. The next generation of linac-based free electron lasers will use very short bunches with a large peak current. For such beams, the impedance caused by sub-micron imperfections in the vacuum beam tube may generate an additional energy spread within the bunch. A review of two mechanisms of the roughness impedance is given with the emphasis on the importance of the high-aspect ratio property of the real surface roughness.

INTRODUCTION

The design of modern free-electron lasers based on self-amplified spontaneous emission (SASE) calls for intense short electron beams with a small relative energy spread [1,2]. For example, in the Linear Coherent Light Source (LCLS) at SLAC the peak current is 3.4 kA, the rms bunch length is 20 microns, and the energy spread is less than 0.1 %. A combination of a high current and a short bunch length raises a concern that the induced wakefields may increase the energy spread beyond the tolerable level. It has been pointed out in [3,4] that a major source of wakefields might be the roughness of the surface of the beam pipe in the FEL undulator.

In the first model of wakefields [3] the roughness was simulated by a collection of bumps of a given shape randomly distributed over a smooth surface. If the bump dimensions are small compared to the bunch length, the impedance in this model is purely inductive. For such simple shapes of the bumps as hemispheres or cubes, the model predicts relatively large impedance and results in severe tolerances on the level of roughness.

A more realistic model of roughness effects was developed in Ref. [5]. In this model, the rough surface is considered as a terrain with a slowly varying slope. As was shown in direct measurements of the surface roughness with Atomic Force Microscope [6], this representation of the roughness is adequate to the real surface of the prototype pipe for the LCLS undulator. In the limit when the bunch length is larger than the correlation length of the roughness, the impedance in this model is also inductive, however the tolerance on the rms height of the surface roughness are much looser than predicted in [3].

CP581, *Physics of, and Science with, the X-Ray Free-Electron Laser,* edited by S. Chattopadhyay et al.
2001 American Institute of Physics 0-7354-0022-9

In yet another approach [4], the roughness wakefield was associated with the excitation of a resonant mode whose phase velocity is equal to the speed of light. The existence of such modes in a round pipe with periodically corrugated walls was studied theoretically in Ref. [7]. In the case when the typical depth of the wall perturbations is comparable to the period, it was shown the the loss factor of such modes reaches the theoretically maximal value for the resonant wakefield. However, as was shown in [8], when the height of the periodic wall corrugations becomes smaller than the period, the loss factor for the mode rapidly decreases. We believe that the latter model is more appropriate for the real roughness of a well finished metal surface.

In this paper we will try to present the results of the latest studies of the roughness impedance, with the emphasis on the realistic modeling of the roughness surfaces.

HOW DOES A ROUGH SURFACE LOOK LIKE?

A naive idea of a rough surface as a microscopic mountain country with sharp peaks and deep canyons does not correspond to reality. A metal surface with a good finish more resembles the water surface of a swimming pool in quiet weather. Pictures of scanned surfaces for different type of machining can be found in surface metrology books [9,10]. Most of them are characterized by that the typical peak-to-valley height h of the roughness is much smaller than the spacing between the crests g. The *aspect ratio* g/h can easily exceed a hundred for smooth surfaces. Of course, this ratio is only one of the many statistical characteristics of the surface, but as it turns out, it is the most essential feature for understanding the electromagnetic interaction of the electron beam with the surface.

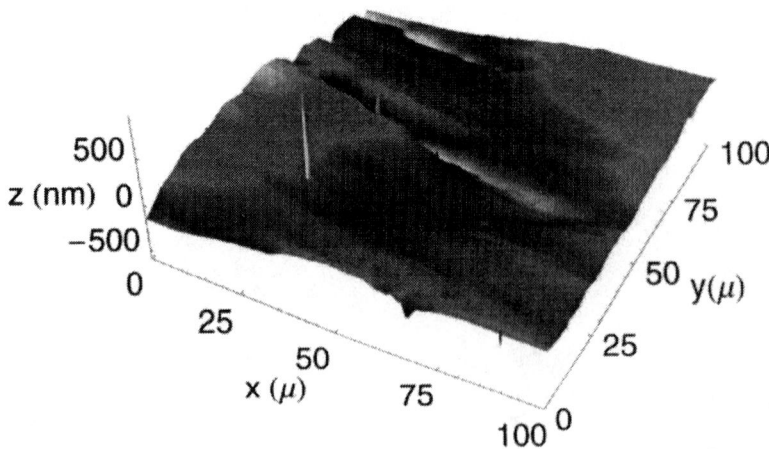

FIGURE 1. A sample surface profile measured with Atomic Force Microscope in Ref. [6]. Note the different scales in the vertical and horizontal directions.

For illustration, we show in Fig. 1 the profile of a surface of a metal pipe measured in Ref. [6]. This pipe is considered as a possible prototype for the vacuum chamber of the LCLS undulator. The rms height of the roughness for this surface is about 100 nm, and the transverse size g, as is seen from the picture, exceeds tens or even a hundred of microns.

SMALL-ANGLE APPROXIMATION IN THE THEORY OF IMPEDANCE

The small ratio h/g implies a small angle θ between the tangent to the surface and the horizontal plane. Using the smallness of this parameter it is possible to develop a perturbation theory of electromagnetic interaction of the beam with the surface based on the so called *small-angle* approximation [5]. This approach extends the earlier treatments [11,12] of an axisymmetric periodic perturbation of the boundary. It also agrees with the more general results of Ref. [13] valid for nonperiodic axisymmetric boundary perturbations.

As follows from Refs. [11,12], for a periodically corrugated wall with the wavelength λ_0 much smaller than the pipe radius b, there exist synchronous modes in the pipe which propagate with the phase velocity equal to the speed of light. The wavelength of these modes is below $2\lambda_0$, so that only a short bunch of length $\sigma_z \lesssim 2\lambda_0$ can efficiently excite these modes. If, on the other hand, the bunch length is larger than λ_0, the excitation of these modes will be exponentially weak. In the roughness problem the parameter g plays the rope of λ_0, and we expect two different regimes depending on whether σ_z is larger or smaller than g.

Long-Bunch Limit, $\sigma_z > g$

In this regime we expect an inductive impedance, because, as explained above, the beam does not lose energy for excitation of the synchronous modes. However, the energy exchange between the head and the tail of the bunch can cause the energy variation along the bunch and may interfere with the lasing.

Let $h(x, z)$ denote the local height of the rough surface as a function of coordinate x in azimuthal direction, and coordinate z along the axis of the pipe (see Fig. 2). The requirement $h \ll g$ can alternatively be expressed as $\theta \approx |\nabla h| \ll 1$. The treatment of Ref. [5] was additionally limited by the assumption that the bunch length is larger than the typical size of the roughness bumps, $\lambda \sim \sigma_z \gg g$. It was found that in this limit the impedance is purely inductive, and the inductance $\mathcal{L}$ per unit length of the pipe is given by the following formula:

$$\mathcal{L} = \frac{Z_0}{2\pi cb} \int_{-\infty}^{\infty} \frac{\kappa_z^2}{\sqrt{\kappa_\theta^2 + \kappa_z^2}} S(\kappa_z, \kappa_\theta) d\kappa_z d\kappa_\theta, \tag{1}$$

143

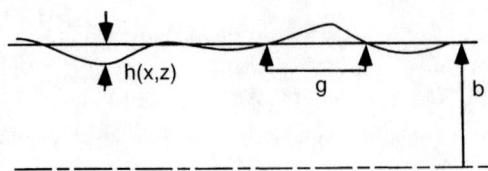

FIGURE 2. The profile of a rough surface. Shown are the height $h(x, z)$, the typical transverse size of the bumps g, and the pipe radius b. Note that the roughness here is not assumed axisymmetric.

where $Z_0 = 4\pi/c = 377$ Ohm, and $S(\kappa_z, \kappa_\theta)$ is the spectrum of the surface profile as a function of wavenumbers k_z and k_θ in the longitudinal and azimuthal directions, respectively. The spectral density S can be defined as a square of the absolute value of Fourier transform of h,

$$S(\kappa_z, \kappa_\theta) = \frac{1}{(2\pi)^2 A} \left| \int_A h(z, x) e^{-i\kappa_z z - i\kappa_\theta x} \, dz dx \right|^2 , \qquad (2)$$

where the integration goes over the surface of a sample of area A. It is assumed that the sample area is large enough so that the characteristic size $\sqrt{A}$ is much smaller than the correlation length g of the roughness.

In Ref. [14] a comparison was done between the small-angle approximation and a previous model of roughness, developed in [3]. It was shown that in the region of mutual applicability both models give the results which, within a numerical factor, agree with each other.

Roughness Measurements

A detailed study of the surface roughness for a prototype of the LCLS undulator pipe using the Atomic Force Microscope was done in Ref. [6]. A high quality Type 316-L stainless steel tubing from the VALEX Corporation with an outer diameter of 6.35 mm and a wall thickness of 0.89 mm with the best commercial finish, A5, was used for the measurements. The samples to be analyzed were cut from this tubing, using an electrical discharge wire cutting process, so as to eliminate damage from mechanical processing. The samples were subsequently cleaned chemically to remove particles adhering to the surface from the cutting process, which used a brass wire.

The measured profiles were Fourier transformed and the inductance $\mathcal{L}$ per unit length of the pipe was calculated using Eq. (1). Because this inductance is inversely proportional to the pipe radius b, a convenient quantity is the product $\mathcal{L}b$ which does not depend on the pipe radius and characterizes the intrinsic properties of the surface. The computed value of this quantity was found to be between 3×10^{-4} pH and 5×10^{-4} pH.

Those values should be compared with the impedance budget for the LCLS beam. For the nominal parameters of LCLS: beam charge 1 nC, $\sigma_z = 20$ μm, undulator length 112 m, and assuming the final beam energy $E = 14.3$ GeV, one finds that the requirement of the relative energy spread $\delta E_{\mathrm{rms}}/E$ generated by the wake be less than 0.05% gives the tolerance $\mathcal{L} < 1.6$ pH/m. For the vacuum pipe radius $b = 2.5$ mm the tolerance on the product $\mathcal{L}b$ is $(\mathcal{L}b)_{\mathrm{tol}} = 4 \times 10^{-3}$ pH. We see that the measured value of the impedance is almost an order of magnitude smaller than the tolerance.

We have to emphasize here that the above results are based on two assumptions that are not completely fulfilled for the LCLS. First, a Gaussian beam distribution was assumed. As detailed simulations show [1], for the LCLS the bunch shape more resembles a rectangular than a Gaussian shape. Second, Eq. (1) used for the calculation of the inductance, was derived in the limit $\sigma_z \gg g$, which, as roughness measurements indicate, is not satisfied. We will show however, in the next section, that using Eq. (1) in the regime of very short bunches, $\sigma_z < g$, overestimates the impedance, and Eq. (1) can be considered as an upper limit for the real impedance of the roughness.

Arbitrary Bunch Length σ_z

Based on the derivation given in Ref. [5] we will calculate here the wakefield of the roughness which is valid for arbitrary relation between σ_z and g. The corresponding impedance can be used even for large frequencies, when $\lambda < g$. For simplicity, we limit our consideration by the case where the pipe wall has a sinusoidal corrugation, as shown in Fig. 3. The amplitude h_0 of the corrugation is assumed much smaller than the period, $h_0 \kappa \ll 1$, which is a requirement of the small-angle approximation. Such a corrugation qualitatively simulates a rough surface with parameter $g \sim \kappa^{-1}$ and the rms height of the bumps of the order of h_0.

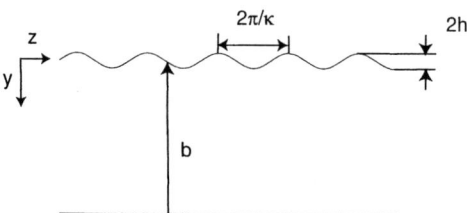

FIGURE 3. A pipe with a sinusoidal corrugation of the wall. The amplitude of the corrugation is h_0, and the period is equal to $2\pi/\kappa$.

The actual derivation of the wake is presented in the Appendix. For the point charge, the longitudinal wake is:

$$w(s) = \frac{h_0^2 \kappa^3}{b} f(\kappa s), \qquad (3)$$

where the function f is

$$f(\zeta) = \frac{1}{2\sqrt{\pi}} \frac{\partial}{\partial \zeta} \frac{\cos(\zeta/2) + \sin(\zeta/2)}{\sqrt{\zeta}}. \qquad (4)$$

The plot of the function f is shown in Fig. 4. It has a singularity $\sim \zeta^{-3/2}$ when $s \to 0$.

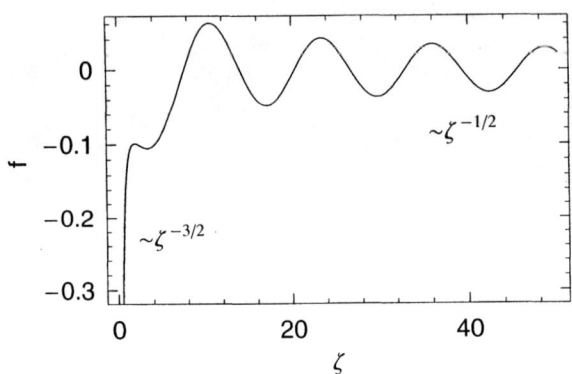

FIGURE 4. The function $f(\zeta)$ for the wake of the sinusoidal corrugation.

To find the wake $W(s)$ for a bunch we need to convolute Eq. (3) with the bunch distribution function $\rho(s)$. For a Gaussian bunch, $\rho(s) = (\sqrt{2\pi}\sigma_z)^{-1} \exp(-s^2/2\sigma_z^2)$, and we have

$$\begin{aligned} W(s) &= \int_s^\infty \rho(s')w(s'-s)ds' \\ &= \frac{h_0^2 \kappa^{3/2}}{b\sigma_z^{3/2}} G\left(\frac{s}{\sigma_z}, \kappa\sigma_z\right), \end{aligned} \qquad (5)$$

where the function G for different values of parameters $\kappa\sigma_z$ is shown in Fig. 5.
In the limit of large and small values of $\kappa\sigma_z$ the wake $W(s)$ scales as

$$\begin{aligned} W(s) &\sim \frac{h_0^2 \kappa}{b\sigma_z^2}, & \sigma_z\kappa \gg 1, \\ &\sim \frac{h_0^2 \kappa^{3/2}}{b\sigma_z^{3/2}}, & \sigma_z\kappa \ll 1. \end{aligned} \qquad (6)$$

We see from these estimates, that when we use long-bunch approximation ($\sigma_z\kappa \gg 1$) in the regime where $\kappa\sigma_z < 1$, we overestimate the wake by a factor of $(\sigma_z\kappa)^{-1/2} \sim$

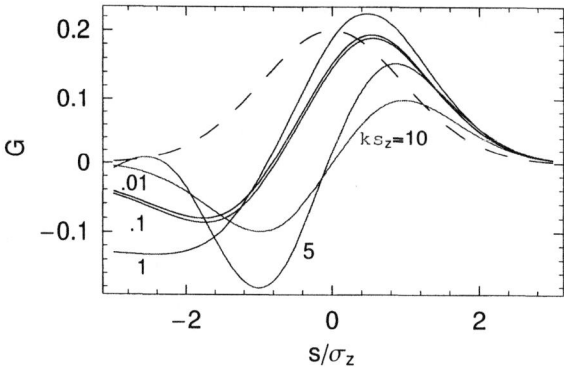

FIGURE 5. The function G for a different values of the parameter $\kappa \sigma_z$ (indicated by numbers on the plot). The dashed curve shows the Gaussian distribution of the beam.

$(g/\sigma_z)^{1/2}$. For this reason, as pointed out at the end of the previous section, the result of Ref. [6] should be considered as an upper boundary for the roughness impedance.

Using Eq. (3) we can also calculate the wake for a rectangular bunch shape, $\rho(s) = 1/l_z$ for $0 < s < l_z$. The result of such calculations for the LCLS is shown in Fig. 6. The parameters used in the calculation are: beam charge 1 nC, $h_0 = 0.28 \ \mu\text{m}$ (corresponding to the rms roughness of 0.2 μm), $g = 2\pi/\kappa = 100 \ \mu\text{m}$, $L = 112$ m, $E = 14.3$ GeV, $b = 2.5$ mm. The averages energy loss for the

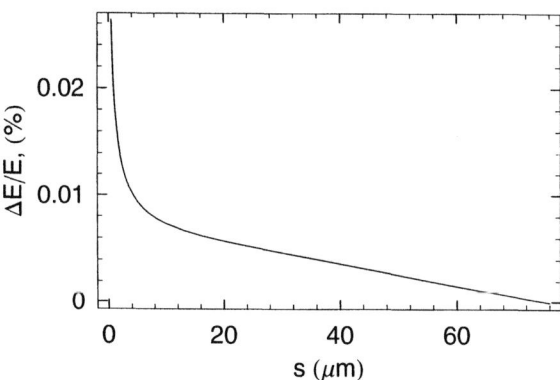

FIGURE 6. The relative energy loss of the LCLS beam at the end of the undulator as a function of position within the bunch.

distribution shown in Fig. 6 is $4.5 \cdot 10^{-3}$ % and the rms energy spread is $2 \cdot 10^{-3}$ %.

147

SYNCHRONOUS MODE

In addition to the mechanism of the wake generation described in the previous section involving interaction with short-wavelength waves, $\lambda \lesssim g$, there is another contribution to the wake which was first pointed out by A. Novokhatski and A. Mosnier [4]. It comes from a relatively low-frequency synchronous mode with $\lambda \gg g$. At first glance, the existence of such mode seems to contradict to the results of Refs. [11,12] which predict that all synchronous modes in a pipe with periodically corrugated surface with the period of corrugation $2\pi/\kappa$ have wavelengths $\lambda < 2/\kappa$. The answer to this apparent contradiction is that this mode arises in the regime where the perturbation theory of [11,12] is not applicable. As we will show below, in the limit when the amplitude of the corrugation tends to zero, the frequency of this mode increases and approaches the value predicted by the perturbation theory. The contribution of the mode to the wake in this limit becomes negligibly small.

It is interesting to note, that earlier a low-frequency mode in a periodically corrugated waveguide was observed in computer simulations in [15], and also studied theoretically in [16].

Rectangular Corrugations of the Wall

The properties of the synchronous mode in the case of rectangular corrugation of the wall were studied in Ref. [7]. In this paper, the wall roughness was modeled by axisymmetric periodic steps on the surface of height δ, width g, and period p. All three parameters were assumed much smaller than the pipe radius b. The model gives for the frequency ω_0 of the mode

$$\omega_0 = c\sqrt{\frac{2p}{\delta bg}}, \tag{7}$$

and for the longitudinal wakefunction of the point charge

$$w(s) = \frac{Z_0 c}{\pi b^2} \cos(\omega_0 s/c). \tag{8}$$

Surprisingly, the amplitude of the wake in this approximation does not depend on the roughness properties at all. These results however are valid if $kp \ll 1$. We see from Eq. (7) that when δ becomes very small, the parameter k increases and eventually kp becomes comparable to unity. Hence, this model becomes invalid in the limit $\delta \to 0$.

The results of computer simulations that confirm the predictions of this model can be found in Ref. [17,18].

Shallow corrugations

To take into account the effect of the shallowness of the roughness a different model was developed in Ref. [8]. In this model the roughness was treated as a sinusoidal perturbation of the wall shown in Fig. 3 with $h_0\kappa \ll 1$. It was found that, indeed, under certain conditions, a low-frequency synchronous mode with $\lambda\kappa \gg 1$ can propagate in this system. The longitudinal wake generated by this mode is given by

$$w(s) = \frac{2Z_0 c}{\pi b^2} U \cos(\omega_0 s/c).$$ (9)

where the dimensionless factor U and the frequency of the mode ω_0 depend on the parameter $r \equiv h_0 \sqrt{b\kappa^3}/2$. The plot of these two functions is shown in Fig. 7. In

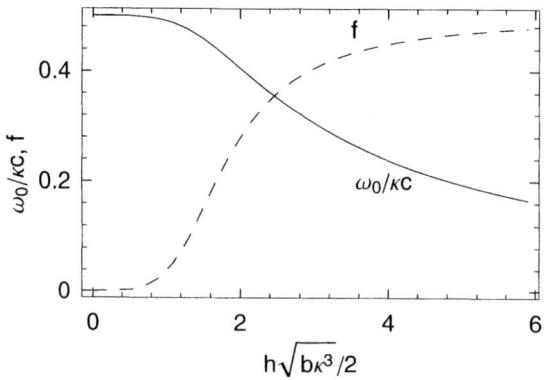

FIGURE 7. Synchronous mode dispersion relation.

the limit $h_0 \to 0$ the frequency ω_0 tends to $\kappa c/2$, and $U \approx r^4/32$. For large values of r, $\omega_0 \approx 2c/h\sqrt{b\kappa}$ and U approaches $1/2$. Interestingly, in this limit we find the amplitude of the wake equal to that in Eq. (7).

Let us estimate the wake for realistic parameters of roughness: $h_0 = 0.28 \ \mu$m (corresponding to the RMS roughness of 0.2 μm), $g = 2\pi/\kappa = 100 \ \mu$m, and $b = 2.5$ mm. We find $r = h_0 \sqrt{b\kappa^3}/2 = 0.11$. The corresponding loss-factor parameter is

$$U \approx 4.5 \cdot 10^{-6},$$ (10)

which indicates that the effect of the wake in this regime will be negligibly small.

CONCLUSIONS

We want to emphasize here that the wakefield generated by the roughness is very sensitive to the geometry of the surface profile. The previous models did not take into account that the real roughness is typically characterized by the large aspect ratio — the ratio of the characteristic size along the surface (correlation length) and the typical height of the bumps. They overestimated the impedance and lead to the very tight tolerances for the surface smoothness.

The latest models of the roughness predict much smaller impedance. Typical numbers that seem safe for the LCLS undulator are: height ~ 100 nm with $g \sim 100$ μm. The surface measurements [6] shows that these parameters are reasonable for a good surface finish.

ACKNOWLEDGEMENTS

The author is thankful to K. Bane and A. Novokhatsky for useful discussions.

This work was supported by Department of Energy contract DE-AC03-76SF00515.

APPENDIX

We write the longitudinal impedance due to roughness as a sum of Eqs. (36) and (43) from Ref. [5]:

$$Z_l(\omega) = -\frac{8k\pi i}{cb^2} \sum_{n,m} \left[1 + \frac{n^2}{(\nu_{n,m}^2 - n^2)} \right] \int_{-\infty}^{\infty} d\lambda \frac{\lambda^2 |\hat{s}_n(\lambda)|^2}{(k + \lambda^2) - (k_{n,m} + i0)^2} , \qquad \text{(A1)}$$

where $k_{n,m} = \sqrt{k^2 - \nu_{n,m}^2/b^2}$, $\nu_{n,m}$ is the mth root of the derivative of the Bessel function J_n', and $\hat{s}_n(\lambda)$ is the Fourier transform of the roughness profile,

$$\hat{s}_n(\lambda) = \frac{1}{(2\pi)^2} \int_{-\infty}^{\infty} dk \int_0^{2\pi} d\theta \, h(z,\theta) e^{i\lambda z + in\theta} . \qquad \text{(A2)}$$

In the limit of high frequency, $\omega \gg c/b$, which was used in the derivation of Eq. (A1), the indices $n, m \gg 1$, and $\nu_{n,m} \approx \mu_{n,m}$ where $\mu_{n,m}$ is the mth root of the Bessel function J_n. Furthermore, in this limit $\nu_{n,m} \approx nf(m/n)$, where the function $f(x)$ is defined by the equation $\pi x = \sqrt{f^2 - 1} - \arccos f^{-1}$. The summation over m and n in Eq. (A1) can be substituted for integration. Introducing new integration variables $k_x = n/b$, $k_z = \lambda$, and f casts the above equation to the following

$$Z_l(\omega) = -\frac{8ki}{cb^2} \int dn \, df \frac{f}{\sqrt{f^2 - 1}}$$
$$\times \int_{-\infty}^{\infty} d\lambda \frac{n\lambda^2 |\hat{s}_n(\lambda)|^2}{(k + \lambda - k_{n,m} - i0)(k + \lambda + k_{n,m} + i0)} . \qquad \text{(A3)}$$

Using the relation

$$\frac{1}{x + i0} = \mathcal{P}\frac{1}{x} - i\pi\delta(x), \tag{A4}$$

where $\mathcal{P}$ stands for the principal part of the integral, we can write the real part of the impedance as

$$\text{Re } Z_l(\omega) = \frac{4\pi k}{cb^2} \int dk_z \, dk_x \, df \, \frac{f k_x k_z^2 |\hat{s}(k_x, k_z)|^2}{\sqrt{f^2 - 1}\sqrt{k^2 - f^2 k_x^2}}$$
$$\times \left[\delta(k + k_z - \sqrt{k^2 - f^2 k_x^2}) + \delta(k + k_z + \sqrt{k^2 - f^2 k_x^2}) \right], \tag{A5}$$

where we now use the notation $\hat{s}(k_x, k_z)$ for $\hat{s}_n(\lambda)b$. Performing integration over k_x and f gives the following result

$$\text{Re } Z_l(\omega) = \frac{4\pi k}{cb^2} \int dk_z \, dk_x \frac{k_z^2 |\hat{s}(k_x, k_z)|^2}{\sqrt{-2kk_z - k_z^2 - k_x^2}}. \tag{A6}$$

In this integral the integration goes over the negative values of k_z such that the expression under the square root is positive. To find the wakefield we use the relation

$$w_l(s) = \frac{2}{\pi} \int_0^\infty \text{Re } Z_l(\omega) \cos\left(\frac{\omega s}{c}\right) d\omega \tag{A7}$$

which gives the following result

$$w_l(s) = \frac{4\sqrt{\pi}}{b^2} \int dk_z \, dk_x |k_z|^{3/2} |\hat{s}(k_x, k_z)|^2 \tilde{w}(k_x, k_z, s) \tag{A8}$$

where

$$\tilde{w} = \frac{\partial}{\partial s} \frac{1}{\sqrt{s}} (\cos qs + \sin qs)$$
$$= \frac{1}{2s^{3/2}} \left[(2qs - 1)\cos qs - (2qs + 1)\sin qs \right], \tag{A9}$$

with $q = (k_x^2 + k_z^2)/2|k_z|$.

It is easy to show that for a sinusoidal perturbation of the surface pipe $h = h_0 \cos \kappa z$ the corresponding spectrum is

$$|\hat{s}(k_x, k_z)|^2 = \frac{h_0^2 Lb}{8\pi} \delta(\kappa + k_z)\delta(k_x). \tag{A10}$$

Putting Eq. (A10) into Eq. (A8) and performing integration gives Eq. (3).

REFERENCES

1. *Linac Coherent Light Source (LCLS) Design Study Report: The LCSL Design Study Group*, Report SLAC-R-521, SLAC (1998).
2. *A VUV Free Electron Laser at the TESLA Test Facility Linac – Conceptual Design Report*, Report TESLA-FEL 95-03, DESY Hamburg (1995).
3. K. L. F. Bane, C. K. Ng, and A. W. Chao, *Estimate of the impedance due to wall surface roughness*, Report SLAC-PUB-7514, SLAC (1997).
4. A. Novokhatski and A. Mosnier, in *Proceedings of the 1997 Particle Accelerator Conference* (IEEE, Piscataway, NJ, 1997), pp. 1661–1663.
5. G. V. Stupakov, Phys. Rev. ST Accel. Beams **1**, 064401 (1998).
6. G. Stupakov, R. E. Thomson, D. Walz, and R. Carr, Phys. Rev. ST Accel. Beams **2**, 060701 (1999).
7. K. L. F. Bane and A. Novokhatskii, *The Resonator impedance model of surface roughness applied to the LCLS parameters*, Tech. Rep. SLAC-AP-117, SLAC (March 1999).
8. G. V. Stupakov, in T. Roser and S. Y. Zhang, eds., *Workshop on Instabilities of High Intensity Hadron Beams in Rings* (American Institute of Physics, New York, 1999), no. 496 in AIP Conference Proceedings, pp. 341–350.
9. D. J. Whitehouse, *Handbook of Surface Metrology* (IOP Publishing, 1994).
10. K. J. Stout, *Atlas of Machined Surfaces* (Chapman and Hall, 1990).
11. M. Chatard-Moulin and A. Papiernik, IEEE Trans. Nucl. Sci. **26**, 3523 (1979).
12. S. Krinsky, in W. S. Newman, ed., *Proc. International Conference on High-Energy Accelerators, Geneva, 1980*, CERN, European Lab. for Particle Physics (Birkhäuser Verlag, Basel, Switzerland, 1980), p. 576.
13. R. L. Warnock, *An Integro-Algebraic Equation for High Frequency Wake Fields in a Tube with Smoothly Varying Radius*, Report SLAC-PUB-6038, SLAC (1993).
14. K. L. F. Bane and G. V. Stupakov, *Wake of a rough beam wall surface*, Tech. Rep. SLAC-PUB-8023, SLAC (December 1998), presented at International Computational Accelerator Physics Conference (ICAP 98), Monterey, CA, 14-18 Sep 1998.
15. K. L. Bane and R. D. Ruth, *Bellows Wake Fields and Transverse Single Bunch Instabilities in the SSC*, Tech. Rep. SLAC/AP-45, Stanford Linear Accelerator Center, Stanford, CA, USA (October 1985).
16. V. I. Balbekov, *Calculation of the Corrugated Vacuum Chamber Impedance, (in Russian)*, Tech. Rep. IFVE-85-128, Inst. for High Energy Physics, Protvino, Russia (1985).
17. M. Dohlus, H. Schlarb, R. Wanzenberg, R. Lorenz, and T. Kamps, *Estimation of longitudinal wakefield effects in the TESLA-TTF FEL undulator beam pipe and diagnostic section*, Tech. Rep. DESY-TESLA-FEL-98-02, Deutsches Elektronen-Synchrotron, Hamburg, Germany (March 1998).
18. A. Novokhatsky, M. Timm, and T. Weiland, *Single bunch energy spread in the TESLA cryomodule*, Tech. Rep. DESY-TESLA-99-16, Deutsches Elektronen-Synchrotron, Hamburg, Germany (March 1999).

X-ray FEL with a meV Bandwidth

E.L. Saldin, E.A. Schneidmiller, Yu.V. Shvyd'ko*, M.V. Yurkov†

Deutsches Elektronen-Synchrotron (DESY), Notkestrasse 85, D-22607 Hamburg, Germany
* II. Institut für Experimentalphysik, Universität Hamburg, D-22761 Hamburg, Germany
† Joint Institute for Nuclear Research, Dubna, 141980 Moscow Region, Russia

Abstract. A new design for a single pass X-ray Self-Amplified Spontaneous Emission (SASE) FEL was proposed in [1] and named two-stage SASE FEL. The scheme consists of two undulators and an X-ray monochromator located between them. For the Angström wavelength range the monochromator could be realized using Bragg reflections from crystals. Proposed scheme of monochromator is illustrated for the 14.4 keV X-ray SASE FEL being developed in the framework of the TESLA linear collider project. The spectral bandwidth of the radiation from the two-stage SASE FEL (20 meV) is defined by the finite duration of the electron pulse. The shot-to-shot fluctuations of energy spectral density are dramatically reduced in comparison with the 100 % fluctuations in a SASE FEL. The peak and average brilliance are by three orders of magnitude higher than the values which could be reached by a conventional X-ray SASE FEL.

I INTRODUCTION

A single pass X-ray SASE FEL [2,3] can be modified as proposed in [1] to reduce significantly the bandwidth and the fluctuations of the output radiation. The modified scheme consists of two undulators and an X-ray monochromator located between them. The first undulator operates in the linear regime of amplification starting from noise and the output radiation has the usual SASE properties. After the first undulator the electron beam is guided through a bypass and the X-ray beam enters the monochromator which selects a narrow band of radiation. At the entrance of the second undulator the monochromatic X-ray beam is combined with the electron beam and is amplified up to the saturation level.

The electron micro-bunching induced in the first undulator should be destroyed prior to its arrival at the second one. This is achieved due to the natural energy spread of the electron beam guided through the bypass. At the entrance of the second undulator the radiation power from the monochromator dominates significantly over the shot noise and the residual electron bunching, and input signal bandwidth is small with respect to the FEL amplifier bandwidth.

The monochromatization of the radiation is performed at a relatively low level of

CP581, *Physics of, and Science with, the X-Ray Free-Electron Laser,* edited by S. Chattopadhyay et al.
© 2001 American Institute of Physics 0-7354-0022-9/01/$18.00

radiation power which allows one to use conventional X-ray optical elements for the monochromator design. X-ray grating techniques can be used successfully down to wavelengths of several Å and at shorter wavelengths crystal monochromators could be used.

The proposed two-stage scheme possesses two significant advantages. First, it opens a perspective to achieve monochromaticity of the output radiation close to the limit given by the finite duration of the electron pulse and to increase the brilliance of the SASE FEL. Second, shot-to-shot fluctuations of the energy spectral density could be reduced from 100 % to less than 10 % when the second undulator section operates at saturation. Since it is a single bunch scheme, it does not require any special time diagram of the accelerator operation.

The conditions that are necessary and sufficient for the effective operation of a two-stage SASE FEL were discussed in [1] and can be summarized as follows

$$P_{\text{in}}^{(2)}/P_{\text{shot}} = G^{(1)} R_{\text{m}} (\delta\lambda/\lambda)_{\text{m}}/(\delta\lambda/\lambda)_{\text{SASE}} \gg 1 \ , \tag{1}$$

$$\lambda/\pi\sigma_{\text{z}} < (\delta\lambda/\lambda)_{\text{m}} \ll (\delta\lambda/\lambda)_{\text{SASE}} \ , \tag{2}$$

$$G^{(1)} \ll G_{\text{sat}}(\text{SASE}) \ . \tag{3}$$

Here $P_{\text{in}}^{(2)}$ is the input radiation power at the entrance to the second undulator, P_{shot} is the effective power of shot noise, $G^{(1)}$ is the power gain in the first undulator, R_{m} is the integral reflection coefficient of the mirrors and the dispersive elements of the monochromator, $(\delta\lambda/\lambda)_{\text{m}}$ is the resolution of the monochromator, $(\delta\lambda/\lambda)_{\text{SASE}}$ is the radiation bandwidth of the SASE FEL at the exit of the first undulator, σ_{z} is the rms length of the electron bunch, and $G_{\text{sat}}(\text{SASE})$ is the power gain of SASE FEL at saturation.

TABLE 1. Parameters of the electron beam and the undulators

Electron beam	
Energy, $\mathcal{E}_0$	25 GeV
Peak current, I_0	5 kA
rms bunch length, σ_{z}	23 μm
Normalized rms emittance, ϵ_{n}	1.6π mm mrad
rms energy spread (entrance)	2.5 MeV
External β-function	45 m
Bunch separation	93 ns
Number of bunches per train	11315
Repetition rate	5 Hz
Undulator	
Type	Planar
Period, λ_{w}	4.5 cm
Peak magnetic field, H_{w}	9.5 kGs

An application of such a two-stage scheme to 6 nm SASE FEL at the TESLA Test Facility at DESY [4] was discussed in [1,5]. Now it is funded and is expected to be the main option for operation of the user facility. In this paper we consider a possible design of the two-stage FEL operating in the Angström range for the X-ray laboratory integrated into the TESLA linear collider project [6]. To be specific we consider the FEL optimized for the 14.4 keV X-rays (0.86 Å). Special interest for the 14.4 keV X-rays is due to additional possibilities which the powerful and diverse nuclear resonance scattering techniques with the highly monochromatic 14.4 Mössbauer radiation [7] open for studies of structure and dynamics of solids, biological molecules, etc.

II PARAMETERS OF THE TWO-STAGE X-RAY FEL

Main parameters of the electron beam and the undulators are presented in Table 1 and coincide with those of the usual SASE FEL at 14.4 keV being designed for the TESLA X-ray laboratory. The SASE FEL bandwidth at the exit of the first stage is about 7×10^{-4} and weakly depends on the gain. We require the monochromator FWHM bandwidth to be about 20 meV, or 1.4×10^{-6} (see (2)). The integral reflection coefficient of all the crystals of the monochromator is expected to be in a range of 0.3 - 0.5. Requiring the excess of the input radiation power at the entrance to the second undulator $P_{in}^{(2)}$ over the effective power of the shot noise P_{shot} to be two orders of magnitude (see (1) and [5]) we end up with a required

TABLE 2. Parameters of the first and the second stages

1st stage	
Wavelength, λ	0.860 Å
Effective power of shot noise, P_{shot}	5 kW
Length of undulator, $L_w^{(1)}$	140 m
FWHM bandwidth , $(\delta\lambda/\lambda)_{SASE}$	7×10^{-4}
Radiation spot size (FWHM)	50 μm
Angular divergence (FWHM)	1 μrad
Peak power	0.75 GW
Average power	4 W
2nd stage	
Input power, $P_{in}^{(2)}$	0.5 MW
Length of undulator, $L_w^{(2)}$	170 m
FWHM bandwidth , $\delta\lambda/\lambda$	1.4×10^{-6}
Angular divergence (FWHM)	0.7 μrad
Radiation spot size (FWHM)	110 μm
Peak power	20 GW
Average power	110 W
Peak brilliance*	3×10^{36}
Average brilliance*	2×10^{28}

*In units of Phot./(sec×mrad^2×mm^2×0.1 % bandw.)

TABLE 3. Parameters of the monochromator and the electron beam bypass

Monochromator	
Nominal energy	14.4 keV
Bandwidth	20 meV
Tunability range	2-4 keV
Total reflection coefficient	0.3 - 0.5
Absorbed average power	< 200 mW
Absorbed average power density	< 50 W/mm^2
Electron beam bypass (chicane)	
Total length	40 m
Bending angle of magnets	1.3°
Path lengthening	1 cm

gain of 1.5×10^5 in the first undulator. The SASE FEL gain at the saturation would be about 4×10^6 so that the condition (3) is satisfied.

Parameters of the first and the second stages are presented in Table 2. They have been calculated with the FEL simulation code FAST [8]. When calculating these parameters we have taken into account the growth of energy spread in the electron beam due to the quantum fluctuations of undulator radiation [9,10]. The peak and average brilliance of the X-ray beam at the exit of the second stage are 500 times larger than in the case of the usual SASE FEL. The shot-to-shot fluctuations of the energy spectral density are reduced to the 10% level due to nonlinear stabilization mechanism [5].

The distance between the two undulators is mainly defined by parameters of the electron beam bypass (chicane) that must compensate a path delay of X-rays in the monochromator. The latter is of an order of 1 cm (c.f. Sect. 3). For a bending angle of the chicane magnets of 1.3° the total length of the chicane is about 40 m. The electron beam microbunching is completely destroyed at the end of the bypass due to the uncorrelated energy spread in the beam and reasonable longitudinal dispersion of the chicane [1]. Due to the small angular divergence of the radiation coming out of the first undulator, the focusing of this radiation is not necessary. Indeed, the calculations show that the input coupling factor [11] to the eigenmode in the second undulator decreases by about 30% with respect to the case of optimal focusing. Main parameters of the monochromator and the electron beam bypass are presented in Table 3.

III HIGH ENERGY-RESOLUTION, HIGH HEAT-LOAD, TUNABLE X-RAY MONOCHROMATOR

The main requirements to the X-ray monochromator of the two stage XFEL are
i. degree of monochromatization: $\lambda/\delta\lambda = E/\delta E = 0.7 \times 10^6$;

ii. tunability range: a few keV;

iii. resistance to the high heatload.

To reach the required value of monochromatization alone is not a problem. Nowadays a monochromatization of 10^7 and more is possible. Bragg diffraction is the main tool used for such purposes. For a recent review of the techniques used and achievements in this field see, e.g., [12]. However, the combination of the three requirements renders the realization of such a monochromator not so straightforward.

A   Spectral width of Bragg reflections and tunability range.

Tunability of an X-ray monochromator for a given monochromaticity will be addressed first.

The relative energy width of a Bragg reflection in a thick nonabsorbing crystal (like silicon, diamond, etc) is given in the dynamical theory of diffraction in perfect crystals (see, e.g., [13]) by

$$\frac{\delta E}{E} = \frac{\delta \lambda}{\lambda} = \frac{|\chi_{\mathbf{g}}|}{\sin^2 \theta}.$$

Here θ is the glancing angle of the radiation plane wave to the reflecting atomic planes (hkl) with the interplanar distance d_{hkl} and the related reciprocal vector $\mathbf{g}$ where $|\mathbf{g}| = 2\pi/d_{hkl}$. The relation between the wavelength λ of the reflected x rays and θ is given by the Bragg law $2d_{hkl} \sin \theta = \lambda$.

$$\chi_{\mathbf{g}} = -\frac{r_e \lambda^2}{\pi V} Z f \left(\frac{\sin \theta}{\lambda} \right) \exp \left(-\frac{\langle u^2 \rangle}{\lambda^2} 8\pi^2 \sin^2 \theta \right)$$

is the Fourier component of the electric susceptibility corresponding to the reciprocal vector $\mathbf{g}$. The expression is valid for a single atom crystal. The following notations are used: V is the volume of the crystal unit cell; Z is atomic number; r_e is the classical electron radius; $f(\ldots)$ is the atomic scattering formfactor; and $\exp(\ldots)$ is the Debye-Waller factor with $\langle u^2 \rangle$ as the mean square displacement of atoms in the direction of the scattering vector $\mathbf{g}$ due to thermal vibrations. The combination of the both equations gives

$$\frac{\delta E}{E} = \frac{r_e \lambda_{hkl}^2}{\pi V} Z f \left(\lambda_{hkl}^{-1} \right) \exp \left(-8\pi^2 \frac{\langle u^2 \rangle}{\lambda_{hkl}^2} \right). \tag{4}$$

Here the Bragg wavelength $\lambda_{hkl} = 2d_{hkl}$ is introduced - the largest wavelength of X-rays allowed by the Bragg law to be reflected from the (hkl) atomic planes.

An important and very favorable implication of eq. (4) for our applications is that the relative spectral width for the given Bragg reflection (hkl) is independent

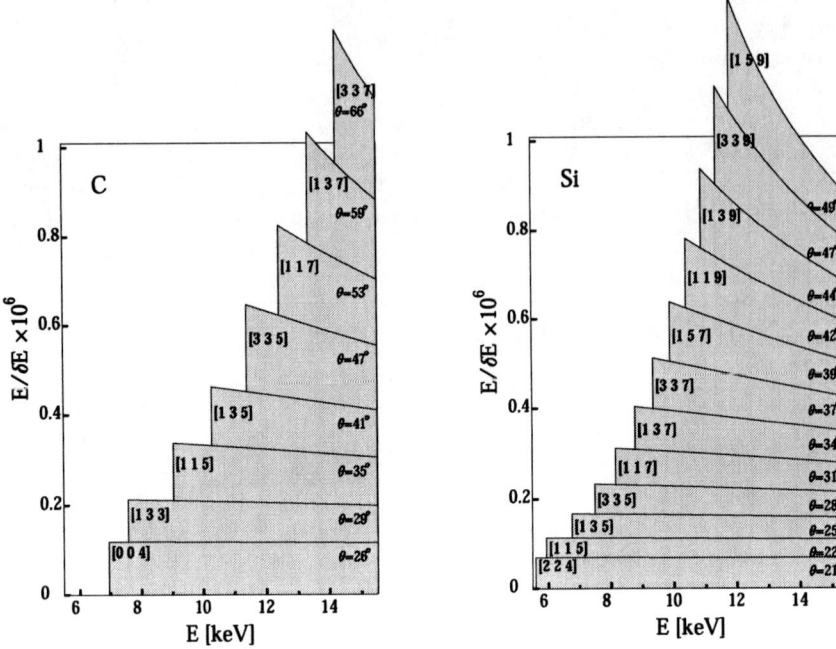

FIGURE 1. Degree of monochromatization $E/\delta E$ of X-rays reflected from the atomic planes (hkl) in diamond (C) and silicon (Si) single crystals at room temperature. The divergence of the incident X-rays is assumed to be 1 μrad. At the right end of each graph the glancing angle θ is given corresponding to the highest X-ray energy $E = 15.5$ keV considered.

of the energy or glancing angle of X-rays and defined merely by properties of the crystal and reflecting atomic planes. In particular it implies that the choice of a crystal, reflecting atomic planes and crystal temperature determines the spectral resolution. Figure 1 shows results of evaluations of the monochromaticity $E/\delta E$ of X-rays reflected from different atomic planes (hkl) in diamond (C) and silicon (Si) single crystals at room temperature. The range of tunability is limited only by the lowest X-ray energy allowed by the Bragg law - the Bragg energy $E_{hkl} = hc/\lambda_{hkl}$.

The 1 μrad divergency (FWHM) of X-rays from the first undulator were also taken into account, which shows up in the decreasing monochromaticity with raising X-ray energy. This occurs when the angular acceptance of Bragg reflections approaches the angular divergence of the incoming beam.

As it is seen from Fig. 1 the number of possible Bragg reflections which provide required monochromaticity and tunability range is rather limited. In case of diamond (C), these are (137) or (117) and equivalent ones. In case of silicon single crystals these are (139) or (339) and equivalent ones.

We are discussing here only silicon and diamond single crystals. There are two reasons for this. Si single crystals are the most perfect crystals available nowadays. This is an important feature which ensures the preservation of the coherent properties of the radiation from the first undulator. Diamond although not so perfect as silicon, nevertheless sufficiently large ≈ 10 mm^2 perfect crystals are available already now [14,15]. The greatest advantage of diamond is its ability to withstand the high heat load due to the extremely high thermal conductivity, low thermal expansion, small X-ray absorption, and high reflectivity.

B   Actual scheme

We have chosen a 4-bounce scheme of the X-ray monochromator as shown in Fig. 2. This solution is advantageous as it allows to keep the direction and the position of the X-ray beam at the exit the same as at the entrance of the monochromator.

This solution is also advantageous since it allows one to use the first two Bragg reflectors as a high-heat load premonochromator, which withdraws the major heat load from the actual high energy-resolution monochromator - the third and the fourth crystals. In front of the first crystal one should install a collimator with a hole of diameter about 100 μm. The spontaneous emission power load on the crystal will be then suppressed to a level much lower than the power of the SASE radiation. In the pre-monochromator part one can use, e.g., diamond crystal plates 100 μm thick reflecting from the atomic planes C(004). Given the crystal is perfect, it reflects 99% of the incident X-rays within a band of 132 meV. Only 5% of the off-band radiation is absorbed, and the rest passes through. The absorbed power is thus 20 times less than the incident one and is about 200 mW. The absorbed power density is about 50 W/mm^2. It is comparable with that at the monochromators of the 3rd generation synchrotron sources [16–18]. The radiation power which reaches the high resolution monochromator crystals is $\simeq 1.3\%$ of the initial value ($\simeq 50$ mW). The latter can be reduced by a factor of two if to use the reflection C(133)

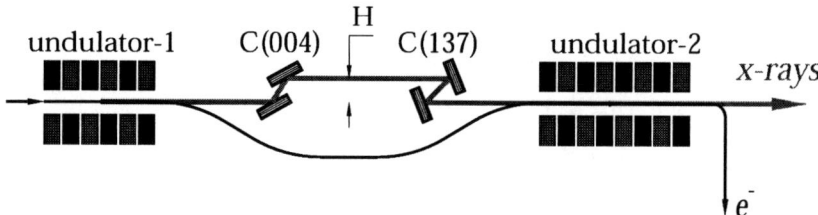

FIGURE 2. The two-stage XFEL with the 4-bounce X-ray monochromator: C(004) $\times$ C(004) $-$ C(137) $\times$ C(137). Additional path length acquired by the X-rays in the monochromator is $\delta L = 3.0H$. Alternative realization is C(004) $\times$ C(004) $-$ Si(139) $\times$ Si(139) with silicon crystals as a high-energy resolution monochromator. The additional path length is $\delta L = 1.7H$.

with a bandwidth of 75 meV.

The final monochromatization to the required level takes place by a high-index reflection from the third and the forth crystals. The required monochromatization $E/\delta E = 0.7 \times 10^6$ of the 14.4 keV X-rays can be achieved, according to Fig. 1, by only a very limited number of Bragg reflections. The final choice of the reflection should be dictated by the requirements of tunability and heatload. The Bragg reflections in diamond have a smaller tunability range. On the other hand they have higher reflectivity and angular acceptance. Fine adjustment of the angular acceptance and the energy bandwidth can be performed by using asymmetric Bragg reflections.

An important technical issue is a path delay (with respect to the straight path) which the X-ray pulse acquires in the monochromator. The path delay equals to

$$\delta L = H \left(\tan\theta_{HKL} + \tan\theta_{hkl} \right), \tag{5}$$

where H is the beam shift, θ_{HKL} in the Bragg angle of the reflections in the high heat-load part (the first two crystals), and θ_{hkl} in the Bragg angle of the reflections in the high energy-resolution part of the monochromator (the third and the forth crystals). By varying H one can keep the delay δL constant in the whole tunability range of the monochromator. For the proposed monochromator schemes the actual values of δL are given in Fig. 2 caption and can be about 1 cm.

IV   CONCLUSION

Our analysis shows that the construction of the high-brilliant two-stage SASE FEL in the Angström spectral range is feasible. We have considered an extreme case when the spectral bandwidth is defined by the length of the electron bunch (lower limit in (2)). By increasing the bandwidth one can increase the tunability range and reduce the power density incident on crystals of the monochromator. The final choice of parameters will be dictated by needs of potential users of intense monochromatic X-rays.

ACKNOWLEDGMENTS

We thank G. Materlik, J. Pflüger and J. Rossbach for their interest in this work. We are grateful to J. Feldhaus and J.R. Schneider for many useful discussions on the two-stage SASE FEL.

REFERENCES

1. Feldhaus, J. et al., *Opt. Commun.*, **140**, 341 (1997).
2. Derbenev, Ya.S., Kondratenko, A.M., and Saldin, E.L., *Nucl. Instr. and Methods* **193**, 415 (1982).

3. Murphy, J.B., and Pellegrini, C., *Nucl. Instr. and Methods* **A237**, 159 (1985).

4. Rossbach, J., *Nucl. Instr. and Methods* **A375**, 269 (1996).

5. Saldin, E.L., Schneidmiller, E.A., and Yurkov, M.V., *Nucl. Instr. and Methods* **A445**, 178 (2000).

6. *Conceptual Design of 500 GeV e^+e^- Linear Collider with Integrated X-ray Facility* (Editors Brinkmann, R. et al.) **DESY 1997-048, ECFA 1997-182**, Hamburg (1997).

7. *Nuclear Resonant Scattering of Synchrotron Radiation*, edited by Gerdau, E., and de Waard, H., *Hyperfine Interactions* **123/124** (1999) and **125** (2000).

8. Saldin, E.L., Schneidmiller, E.A., and Yurkov, M.V., *Nucl. Instr. and Methods* **A429**, 233 (1999).

9. Saldin, E.L., Schneidmiller, E.A., and Yurkov, M.V., *Nucl. Instr. and Methods* **A374**, 401 (1996).

10. Saldin, E.L., Schneidmiller, E.A., and Yurkov, M.V., *Nucl. Instr. and Methods* **A381**, 545 (1996).

11. Saldin, E.L., Schneidmiller, E.A., and Yurkov, M.V., *The Physics of Free Electron Lasers*, Springer, Berlin (1999).

12. Toellner, T., *Hyperfine Interactions*, **125**, 3 (2000).

13. Pinsker, Z.G., *Dynamical Scattering of X rays in Crystals*, Springer, Berlin (1978).

14. Ishikawa, T., et al., *Proc. SPIE* **4145** (2000).

15. Freund, A.K., et al., *Nucl. Instr. and Methods A* (2001).

16. Freund, A.K., *Review of Scientific Instruments* **67**, 5 (1996).

17. Mills, D.M., *J. Synchrotron Rad.* **4**, 117 (1997).

18. Bilderback, D.H., et al., *J. Synchrotron Rad.* **7**, 53 (2000).

Development of a Facility for Probing the Structural Dynamics of Materials with Femtosecond X-ray Pulses

B. Faatz, A.A. Fateev*, J. Feldhaus, K. Floettmann,

T. Tschentscher, J. Krzywinski†, J. Pflueger, J. Rossbach,

E.L. Saldin, E.A. Schneidmiller, M.V. Yurkov*

Deutsches Elektronen-Synchrotron (DESY), Notkestrasse 85, D-22607 Hamburg, Germany
* Joint Institute for Nuclear Research, Dubna, 141980 Moscow Region, Russia
†Institute of Physics of the Polish Academy of Sciences, 02688 Warszawa, Poland

Abstract. We propose to use Thomson backscattering of far-infrared (FIR) pulses (100-300 μm wavelength range) by a 500 MeV electron beam to generate femtosecond X-rays at the TESLA Test Facility (TTF) at DESY. Using the parameters of the photocathode rf gun and the magnetic bunch compressors of the TESLA Test Facility (TTF), it is shown that electron pulses of 100-fs (FWHM) duration can be generated. Passing the short electron bunches through an undulator (after the conversion point) can provide a FIR high-power source with laser-like characteristics. On the basis of the TTF parameters we expect to produce X-ray pulses with 100-fs duration, an average brilliance of nearly 10^{13} photons s^{-1} mrad^{-2} mm^{-2} per 0.1% BW at a photon energy 50 keV. The total number of Thomson backscattered photons, produced by a single passage of the electron bunch through the mirror focus, can exceed 10^7 photons/pulse. We also describe the basic ideas for an upgrade to shorter X-ray pulse duration. It is demonstrated that the TTF has the capability of reaching the 10^{12} photons s^{-1} mrad^{-2} mm^{-2} per 0.1% BW brilliance at a ten femtosecond scale pulse duration.

I INTRODUCTION

Understanding the structural dynamics of materials on the fundamental time scale of atomic motion represents an important frontier in condensed matter research because chemical reactions, phase transitions, and surface processes are ultimately driven by the motion of atoms on the time scale of one vibrational period ($\simeq$ 100 fs). Recent efforts at applying X-rays to probe structural dynamics have used a synchrotron source combined with a femtosecond optical laser [1]. Femtosecond synchrotron radiation pulses were generated directly from an electron storage ring ALS. An ultrashort laser pulse was used to modulate the energy of electrons within a 100-femtosecond slice of the stored 30-picosecond electron bunch. The

CP581, *Physics of, and Science with, the X-Ray Free-Electron Laser*, edited by S. Chattopadhyay et al.

TABLE 1. Major electron beam parameters for 100 fs-scale X-ray facility option

Parameter	Value
Beam energy	500 MeV
Transverse normalized emittance	$2\,\pi$ mm-mrad
Longitudinal emittance	$30\,\pi$ keV-mm
Bunch charge	1 nC
Bunch repetition rate	9 MHz
Duty factor	1%

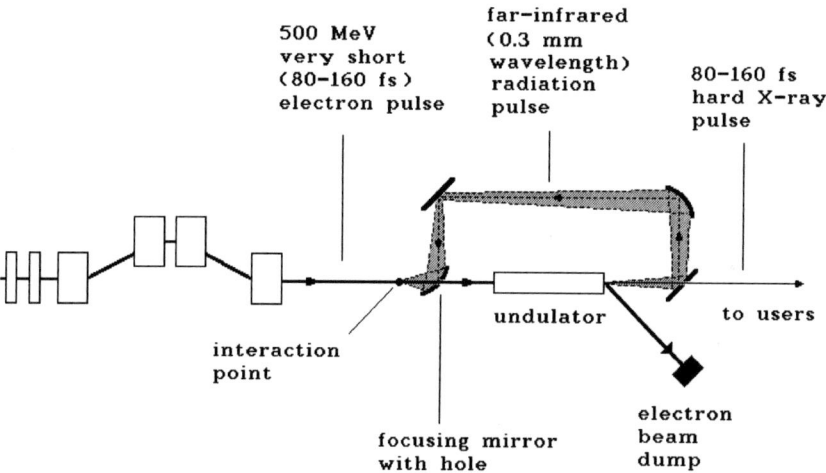

FIGURE 1. Basic scheme of the femtosecond X-ray facility.

energy-modulated electrons were spatially separated from the long bunch and used to generate 300-femtosecond X-ray pulses at a bending magnet beamline. The same technique can be used to generate 100-femtosecond X-ray pulses. On the basis of the parameters of an ALS small-gap undulator and for example, laser pulses of 25 fs and 100 μJ at a repetition of rate 20 kHz, Schoenlein and co-workers expect in the future an average brilliance of 10^{11} photons s^{-1} mrad^{-2} mm^{-2} per 0.1% BW at a photon energy of 2 keV.

Another project of a femtosecond X-ray facility, which is described in detail in this paper, is based on the idea to use Thomson backscattering of a high power far-infrared radiation pulse from a relativistic electron bunch. In our project we use 500 MeV electron bunches from the Tesla Test Facility (TTF) linear accelerator [2] and 200 MW optical pulses from a FIR source based on coherent undulator radiation (see Fig. 1). The basic TTF electron beam parameters are given in Table 1. In a backscattering geometry, the duration of the scattered X-ray burst is determined by the length of the electron bunch. In order to achieve a small bunch length in

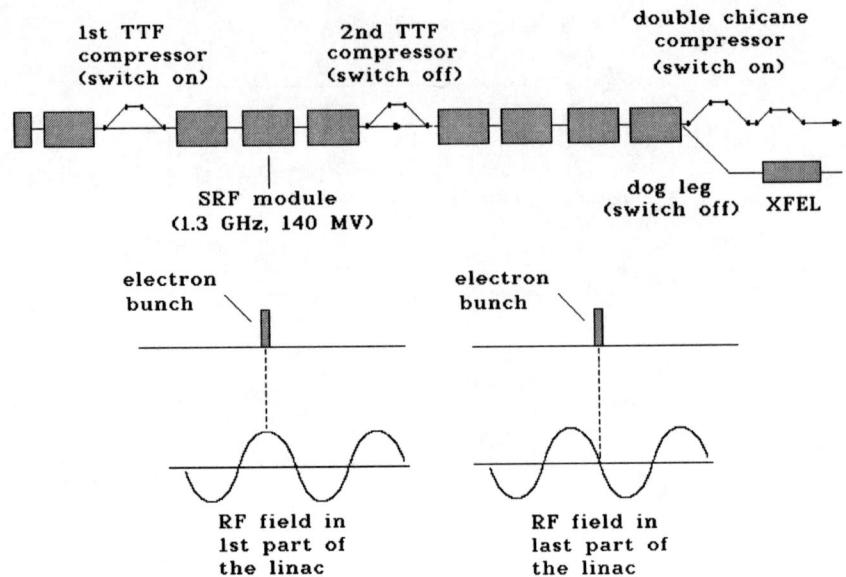

FIGURE 2. Femtosecond X-ray facility compression and acceleration schematic.

the conversion point, the bunch must be compressed in magnetic chicanes. For this project we assume to use a two-stage compressor design (see Fig. 2). The first TTF magnetic chicane compresses the bunch from $\sigma_z \simeq 1.6$ mm to $\sigma_z \simeq 0.4$ mm. After the first bunch compressor the electron bunch is accelerated in the next part of the TTF linac from 150 MeV to 500 MeV energy. Then the electron bunch passes the last part of the accelerating structure (voltage $V = 500$ MV) at 90° crossing phase and the energy spread of the electron beam is increased to about 5 MeV. The electron bunch with such large correlated energy spread can be compressed in special double magnetic chicane down to $\sigma_z \simeq 10$ μm.

II YIELD OF X-RAY PHOTONS

We propose to install a special undulator after the interaction point. Intense, coherent far-infrared undulator radiation can be produced from electron bunches at wavelengths longer than the bunch length. The undulator equation $\omega_0 = 2ck_{\mathrm{w}}\gamma^2 \left[1 + K^2/2\right]^{-1}$ tells us the resonance frequency of radiation as a function of undulator period $\lambda_{\mathrm{w}} = 2\pi/k_{\mathrm{w}}$, undulator parameter K and relativistic factor γ. Note that for radiation within the central cone the relative spectral bandwidth $\Delta\omega/\omega \simeq 1/N_{\mathrm{w}}$, where N_{w} is the number of undulator periods. The energy radiated into the central cone in the case when the resonance wave length much longer that the bunch length is given by ($K^2 \gg 1$): $\Delta\mathcal{E}_{\mathrm{con}} \simeq \pi e^2 \omega_0 N_{\mathrm{e}}^2/c$.

The planar undulator is an inexpensive electromagnetic device with 20 periods, each 60 cm long. At the operation wavelength of the FIR source around 100 μm and an electron beam energy of 500 MeV the peak value of the magnetic field is about 0.5 T. In the case when the number of electron per bunch is about $N_e = 6 \times 10^9$ (1 nC) the FIR source described above provides a train of 6.6 ps long micropulses (the wave advances the electron beam by one wavelength at one undulator period), with 1.5 mJ of optical energy per micropulse radiated into the central cone at 100 μm wavelength. To provide a natural selection of coherent radiation in the central cone ($\theta_{\text{cone}} \simeq 4$ mrad), the radius of the mirror should be equal to 5 cm at a distance of about 10 meters between the exit of the FIR undulator and the mirror.

To obtain an effective conversion of the primary photons into X-ray photons, the far-infrared beam should be focused on the electron beam. This may be performed, for instance, by means of a metal focusing mirror (see Fig. 1). Electrons move along the z axis and pass through the mirror focus. The conditions of optimal focusing are as follow ($l_e \ll l_{\text{opt}}$):

$$\sigma^2_{x,y} \ll [4cF/(\omega a_0)]^2 \;, \quad F^2 \ll a_0^2 l_{\text{opt}}/(2\lambda) \;,$$

where $\sigma_{x,y} = \sqrt{\epsilon_{x,y}\beta_{x,y}}$ is the transverse electron beam size at the conversion point, $\beta_{x,y}$ is the beta function, $\epsilon_{x,y}$ is transverse emittance of the electron beam, F is the focal length of the mirror, a_0 is the radius of the optical beam spot on the mirror, l_e and l_{opt} are the lengths of electron and optical bunch, respectively. The first condition assumes the transverse size of the electron beam at the conversion point to be much less than the FIR beam size. The second condition means that the characteristic axial size of the region with a strong optical field is much less than the length of the optical bunch.

Hard X-ray photons are produced by means of Thomson backscattering of the optical photons by the high energy electrons. Relativistic effects cause the X-ray flux to be strongly peaked in the forward direction; for the 500 MeV electron beam ($\gamma = 10^3$) and 100 μm radiation wavelength used in our project, the Thomson backscattered X-rays are peaked at a maximum energy $\hbar\omega_{\text{x-ray}} = 4\gamma^2\hbar\omega_{\text{opt}} \simeq 50$ keV.

When the conditions of optimal focusing are fulfilled the total number of X-ray photons, produced by a single passage of the electron beam through the mirror focus, is given by the following relation:

$$\Delta N_{\text{x-ray}} = \frac{2W\sigma_{\text{T}}}{\hbar c^2} N_e \;,$$

where σ_{T} is the total Thomson cross section, N_e is the number of electrons in the bunch, and W is the peak power of the optical beam. An important feature of the obtained result is that the number of produced X-ray photons does not depend on the details of the optical field distribution at the mirror surface and is defined by the peak optical beam power only. Taking into account the TTF parameters and

FIR source parameters (see Table 1) and assuming that the conditions of optimal focusing are fulfilled, we obtain a yield of X-ray photons of $\mathrm{d}\,N_{\text{x-ray}}/\mathrm{d}\,t \simeq 2 \times 10^{12}$ s^{-1} .

The quality of the radiation source is described usually by the brilliance defined as the density of photons in the six-dimensional phase space volume:

$$B = \frac{1}{4\pi^2 \sigma_x \sigma_x' \sigma_y \sigma_y'} \left(\lambda \frac{\mathrm{d}^2 N_{\text{ph}}}{\mathrm{d}\,\lambda\,\mathrm{d}\,t} \right) \, .$$

Let us proceed with numerical example for 160 fs (FWHM) electron pulse duration. The main parameters of the electron beam are presented in Table 1. The source of primary photons has the following parameters: wavelength 100 μm, peak power $W \simeq 250$ MW, pulse duration 6.6 ps. Assuming the focus distance of the mirror to be equal to $F \simeq 20$ cm and the radius of the FIR beam at the focusing mirror $a_0 \simeq 10$ cm we find that the condition of optimal focusing will be fulfilled at $\beta \simeq 10$ cm. Under these conditions we calculate an average brilliance of 10^{13} photons s^{-1} mrad^{-2} mm^{-2} per 0.1% BW at a photon energy of 50 keV

In order to produce 80 fs long electron pulses the bunch charge must be reduced to about 0.3 nC. It will be done in stages to avoid coherent synchrotron radiation effects (longitudinal wakefield) limiting the achievable bunch length and beam emittance. On the basis of the parameters of the electron beam with 80-fs pulses and the FIR source we expect an average brilliance of about 10^{12} photons s^{-1} mrad^{-2} mm^{-2} per 0.1% BW at a photon energy of 50 keV.

III   BUNCH COMPRESSORS

The proposed phase bunching system is sketched in Fig. 2. Compressing the bunch will be done in stages to avoid space charge and coherent synchrotron radiation (CSR) effects limiting the achievable bunch length and transverse emittance. The first compression is from 1.6 mm to 400 μm. It consists of a 150 MeV accelerating module followed by the first TTF magnetic chicane generating the R_{56} needed for bunch compression [3]. Calculations show that induced energy spread and emittance dilution should not be a serious limitation in BC1 [3].

After leaving the first bunch compressor the electron bunch of 400μm length is accelerated in the next part of the TTF linac with an on-crest phase from 150 MeV to 500 MeV. For the second compression the required large correlated energy spread in the bunch of 5 MeV is induced by passing the last part of the accelerating structure at 90° crossing phase. Our analysis shows that adequate solution for the BC2 design is a double chicane. Each chicane is twelve meters long and contains four C-type rectangular bending magnets. For compression from 400 μm to 20 μm the parameters of BC2 are practically identical to that designed for LCLS BC2 [4]. The first and second chicanes generate $R_{56} = 3$ cm and 0.3 cm at bend angle $\theta_{B1} = 3.4°$ and $\theta_{B2} = 1.3°$, respectively. Four quadrupoles are placed between the chicanes in locations where the dispersion passes through zero. Since the energy spread generated by CSR is coherent along the bunch, its effect on the transverse emittance can be compensated in a double chicane with optical symmetry to cancel the longitudinal-to-transverse coupling [4].

IV FUTURE POTENTIAL

For the 100-femtosecond X-ray facility described above, we have adopted the TTF design parameters for the photocathode rf gun injector [2]. In order to achieve a 5 μm rms bunch length (40-fs pulse duration) a new mode of operation of the TTF injector is required. The injector should produce 100 pC bunches with a longitudinal emittance of 10 π keV-mm and a normalized transverse emittance of 1 π mm-mrad. In order to reach $B \simeq 10^{13}$ photons s^{-1} mrad^{-2} mm^{-2} per 0.1% BW, the bunch repetition rate (within a macropulse) would have to be increased up to 108 MHz. Such values of bunch charge and repetition rate would keep the mean value of average current during a macropulse and the mean power of the laser system (below 2 W) at the TTF design level. As a result, no modifications are necessary for the rf gun. In this high repetition rate option only the laser system needs hardware modifications.

Analysis of the TTF parameters shows that an extension to bunch lengths shorter than 5 μm rms is also direct and straightforward but needs increasing linear correlated energy spread. Compressing the bunch in our case will be done in stages to avoid RF curvature effect. The proposed solution is to perform compression in three steps: compression at 150 MeV, decompression at 500 MeV before the last part of the linac and compression at 500 MeV at the linac exit. In this scheme the energy spread of the electron beam is increased to about 15 MeV. An electron bunch with such a large correlated energy spread and 30 pC charge per bunch, can be compressed in a special double magnetic chicane down to $\sigma_z \simeq 2$ μm. The average brilliance of such a 10 fs-scale X-ray facility could reach a value of $B \simeq 10^{11}$ photons s^{-1} mrad^{-2} mm^{-2} per 0.1% BW. This requires installation of an additional chicane for decompression and modification of the photoinjector laser system.

The yield of X-ray photons may be increased further more by means of organizing several conversion points. After crossing the first conversion point, the optical beam is directed to an optical delay line and then it is focused at the next electron bunch, etc. Taking into account that the reflection losses of metal focusing mirrors for radiation of 100 μm wavelength are only about 0.5%, we may conclude that each optical bunch can effectively interact with many electron bunches. As a result, a yield of X-ray photons may be increased by a factor of about 10.

ACKNOWLEDGMENTS

We thank D. Trines and J.R. Schneider for interest in this work and support.

REFERENCES

1. Schoenlein, R.W., et al., *Science* **287**, 2237 (2000).
2. Rossbach, J., *Nucl. Instr. and Methods* bf A375, 269 (1996).
3. Limberg, T., et al., *Nucl. Instr. and Methods* bf A375, 322 (1996).
4. *Linac Coherent Light Source (LCLS) Design Study Report* (The LCLS Design Study Group, Stanford Linear Accelerator Center) **SLAC-R-521** (1998).

Diffraction Effects in the SASE FEL

E.L. Saldin, E.A. Schneidmiller, M.V. Yurkov*

Deutsches Elektronen-Synchrotron (DESY), Notkestrasse 85, D-22607 Hamburg, Germany
* Joint Institute for Nuclear Research, Dubna, 141980 Moscow Region, Russia

Abstract. In this paper we present a systematic approach for analytical description of SASE FEL in the linear mode. We calculate the average radiation power, radiation spectrum envelope, angular distribution of the radiation intensity in far zone, degree of transverse coherence etc. Using the results of analytical calculations presented in reduced form, we analyze various features of the SASE FEL in the linear mode. The general result is applied to the special case of an electron beam having Gaussian profile and Gaussian energy distribution. These analytical results can be serve as a primary standard for testing the codes. In this paper we present numerical study of the process of amplification in the SASE FEL using three-dimension time-dependent code FAST. Comparison with analytical results shows that in the high-gain linear limit there is good agreement between the numerical and analytical results. It has been found that even after finishing the transverse mode selection process the degree of transverse coherence of the radiation from SASE FEL visibly differs from unity. This is consequence of the interdependence of the longitudinal and transverse coherence. The SASE FEL has poor longitudinal coherence which develops slowly with the undulator length thus preventing a full transverse coherence.

I INTRODUCTION

A complete description of the SASE FEL can be performed only with three-dimensional (3-D) time-dependent numerical simulation codes. Application of the numerical calculations allows one to describe the general case of the SASE FEL operation, including the case of an arbitrary axial and transverse profile of the electron bunch, the effects of finite pulse duration and nonlinear effects. Since construction of the 3-D time-dependent codes is a rather complicated problem, significant attention should be devoted to the testing the codes. On the other hand, testing the numerical simulation codes would be difficult without the use of analytical results of SASE FEL linear theory as a primary standard. With the design and construction of VUV and X-ray FELs, many 3-D time-dependent codes (GINGER [1], GENESIS [2], FAST [3]) have been developed over the years in order to describe FEL amplifier start-up from shot noise. Nevertheless, it should be emphasized that despite these codes are widely used in the design of X-ray FELs

CP581, *Physics of, and Science with, the X-Ray Free-Electron Laser,* edited by S. Chattopadhyay et al.
© 2001 American Institute of Physics 0-7354-0022-9/01/$18.00

[4–7], there are no comparison between numerical simulation and analytical results of 3-D SASE FEL theory.

From the theoretical point of view the SASE FEL, is a rather complicated object, so it is important to find a model which provides the possibility of an analytical description without loss of essential information about the features of the SASE process. When deriving analytical results we used the model of a long electron bunch with rectangular axial profile of the current. Investigation of the SASE FEL process is preformed with steady-state spectral Green's function connecting the Fourier amplitudes of the output field and the Fourier amplitudes of the input noise signal. Since in the linear regime all the harmonics are amplified independently, we can use the result of steady-state theory for each harmonic and calculate the corresponding Fourier harmonics of output radiation field. In the framework of this model it becomes possible to describe analytically all the statistical properties of the radiation from the SASE FEL.

When the FEL amplifier operates in the steady-state linear regime, the driving electron beam can be considered as an active medium whose properties do not depend on the longitudinal coordinate z. Let us analyze the nature of the self-consistent solution of Maxwell's equations and the Vlasov equation at a fixed frequency ω. The electric field of the wave radiated in the helical undulator may be represented in the complex form:

$$E_x + iE_y = \tilde{E}(z, \vec{r}_\perp) \exp[i\omega(z/c - t)] , \tag{1}$$

At a sufficient distance from the undulator entrance the radiation can be presented as a superposition of the exponentially growing guided modes

$$\tilde{E}(z, \vec{r}_\perp) = \sum_j A_j \Phi_j(\vec{r}_\perp) \exp(\lambda_j z) ,$$

where λ_j and $\Phi_j(\vec{r}_\perp)$ are the eigenvalues and the eigenfunctions of the guiding modes, respectively, and $\mathrm{Re}(\lambda_j) > 0$. The rigorous solution of the eigenvalue problem for an axisymmetric electron beam with stepped profile was obtain in [8]. The model of the FEL amplifier considered in that paper is based on a full three-dimensional description of electromagnetic field, but the electron motion is considered to be one-dimensional. Later the initial-value problem was solved in [9] in framework of the same model of the electron beam. An effects similar to the optical guiding effects occurs in optical fibers. However, unlike guided modes in fiber optics, FEL guided modes are not orthogonal. To solve the initial value problem in the case of arbitrary gradient beam profile, approaches other than direct mode expansion must be used. The first step in this direction was taken by Kim [10], who has applied a method of solution, originally introduced by van Kampen [11]. A Laplace transform method was employed by Krinsky and Yu [12], leading to a Green's function. This Green's function can still be expanded in terms of orthonormal eigenfunctions of the associated two-dimension Schrödinger equation

170

with non-self-adjoint Hamiltonian. In the high-gain limit, the asymptotic representation of the Green's function is found to be dominated by the contribution of the guided modes. The eigenvalue problem for the case of an arbitrary gradient axisymmetric profile is solved by means of the multilayer approximation method [13]. Based on these solutions, complete information on the eigenfunctions and eigenvalues can be extracted, and used, for calculations of the output radiation field. General solution for Green's function [12] gives us input coupling factors A_j.

In this paper we present a systematic approach for calculations of the average radiation power, radiation spectrum envelope, and angular distribution of the radiation intensity in the far zone, and degree of transverse coherence. These analytical results serve as a primary standard for testing the codes. Numerical simulations have been performed with 3-D time-dependent code FAST [3]. Comparison with analytical results shows that in the high-gain linear limit there is a good agreement between the numerical and analytical results.

II ANALYTICAL DESCRIPTION OF THE STEADY-STATE LINEAR REGIME

Let us consider electron beam moving along the z axis in the field of a helical undulator. The magnetic field of the undulator may be written in the complex form: $H_x = H_w \cos(k_w z)$, $H_y = -H_w \sin(k_w z)$. We neglect the transverse variation of the undulator field and assume the electrons move along constrained helical trajectories in parallel with the z axis. The electron rotation angle is considered to be small and the longitudinal electron velocity v_z is close to the velocity of light c. Let us consider a axisymmetric electron beam with gradient profile of the current density. The general form of the transverse distribution of the beam current density of axisymmetric electron beam (in cylindrical coordinates (r, φ, z)) is

$$j_0(r) = I_0 S(r/r_0) \left[2\pi \int_0^\infty r S(r/r_0) \mathrm{d}r \right]^{-1} , \tag{2}$$

where r_0 is the beam profile parameter (typical transverse size of the beam) and I_0 is the beam current. To be specific, we set $S(0) = 1$.

We describe the electron motion using energy-phase variables $P = \mathcal{E} - \mathcal{E}_0$ and $\psi = k_w z + \omega(z/c - t)$, where $\mathcal{E}$ is the kinetic energy of the electron, $\mathcal{E}_0$ is the nominal energy. The evolution of the distribution function of the electron beam is governed by the Vlasov equation. We solve this equation using the perturbation method, so the beam current density is given by:

$$j_z = -j_0(r) + \sum_{n=-\infty}^{n=+\infty} \tilde{j}_1^{(n)}(z, r) \, \mathrm{e}^{-\mathrm{i}n\varphi + \mathrm{i}\psi} + \text{C.C.} ,$$

At a sufficient distance from the undulator entrance the output radiation can be presented as a superposition of the "self-reproducing" field configurations. It is reasonable to represent $\tilde{E}$ as a Fourier series in the angle φ:

$$\tilde{E}(z,r,\varphi) = \sum_{n=-\infty}^{n=+\infty} \tilde{E}^{(n)}(z,r)\,\mathrm{e}^{-\mathrm{i}n\varphi}\,,$$

and write $\tilde{E}^{(n)}$ in dimensionless form

$$\hat{E}^{(n)}(\hat{z},\hat{r}) = \sum_{j} A_j^{(n)} \Phi_{nj}(\hat{r}) \exp(\lambda_j^{(n)}\hat{z})\,. \qquad (3)$$

In this paper we consider the specific, but important practical case of the following initial conditions: - the electron beam is modulated only in density at the undulator entrance; - the field amplitude of the electromagnetic wave $\tilde{E}$ takes the value $\tilde{E}|_{z=0} = 0$ at the undulator entrance. The input coupling factors $A_j^{(n)}$ are given by the expression [10,12]:

$$A_j^{(n)} = 2u_j^{(n)} \int_0^\infty \hat{a}_{\mathrm{ext}}^{(n)}(\hat{r}) S(\hat{r}) \Phi_{nj}(\hat{r})\hat{r}\mathrm{d}\hat{r}\,,$$

where $u_j^{(n)}$ can be written in the following form:

$$u_j^{(n)} = \hat{D}_0(\lambda_j^{(n)}) \left[B \int_0^\infty \Phi_{nj}^2(\hat{r})\hat{r}\mathrm{d}\hat{r} - \left(\frac{\mathrm{d}\hat{D}}{\mathrm{d}p}\right)_{p=\lambda_j^{(n)}} \int_0^\infty \Phi_{nj}^2(\hat{r})S(\hat{r})\hat{r}\mathrm{d}\hat{r} \right]^{-1}\,.$$

Here the following notation is introduced: $C = k_{\mathrm{w}} - \omega/(2c\gamma_z^2)$ is the detuning of the electron with the nominal energy $\mathcal{E}_0$,

$$\theta_{\mathrm{s}} = eH_{\mathrm{w}}/(\mathcal{E}_0 k_{\mathrm{w}}) = K/\gamma\,, \quad \gamma_z^{-2} = \gamma^{-2} + \theta_{\mathrm{s}}^2\,, \quad \gamma = \mathcal{E}_0/(m_e c^2)\,,$$

$\hat{z} = \Gamma z\,, \quad \hat{r} = r/r_0\,, \quad \hat{C} = C/\Gamma$ is the detuning parameter, $B = r_0^2 \Gamma \omega/c$,

$$\Gamma = \left[I_0 \omega^2 \theta_{\mathrm{s}}^2 \left(2I_A c^2 \gamma_z^2 \gamma \int_0^\infty \zeta S(\zeta)\mathrm{d}\zeta \right)^{-1} \right]^{1/2}$$

$I_A \simeq 17$ kA is the Alfven current. The complex amplitudes $\tilde{E}^{(n)}$ and $\hat{E}^{(n)}$ are connected by the relation $\hat{E}^{(n)} = \tilde{E}^{(n)}/E_0\,, \quad E_0 = \rho\mathcal{E}_0\Gamma/(e\theta_{\mathrm{s}})\,, \quad \rho = c\gamma_z^2\Gamma/\omega\,.$ The complex amplitude of the first harmonic of the beam current density $\tilde{j}_1^{(n)}$ is connected with $\hat{a}_1^{(n)}$ by the relation

$$\hat{a}_1^{(n)}(\hat{z},\hat{r}) = -\tilde{j}_1^{(n)}(\hat{z},\hat{r})/j_0(\hat{r})\,, \quad \hat{a}_{\mathrm{ext}}^{(n)}(\hat{r}) = \hat{a}_1^{(n)}(\hat{z},\hat{r})|_{z=0}\,.$$

For a Gaussian energy spread in the electron beam

$$F(\mathcal{E} - \mathcal{E}_0) = (2\pi\langle(\Delta\mathcal{E})^2\rangle)^{-1/2} \exp\left(-\frac{(\mathcal{E} - \mathcal{E}_0)^2}{2\langle(\Delta\mathcal{E})^2\rangle} \right)\,,$$

the functions $\hat{D}(p)$ and $\hat{D}_0(p)$ are given by

$$\begin{pmatrix} \hat{D} \\ \hat{D}_0 \end{pmatrix} = \int_0^\infty \begin{pmatrix} i\xi \\ 1 \end{pmatrix} \exp\left[-\hat{\Lambda}_T^2 \xi^2/2 - (p + i\hat{C})\xi\right] d\xi ,$$

where $\hat{\Lambda}_T^2 = \langle(\Delta\mathcal{E})^2\rangle/(\rho^2\mathcal{E}_0^2)$.

An important characteristic of the FEL amplifier is the output power. In the paraxial approximation the power of the radiation with azimuthal index n, can be written in the following normalized form:

$$\hat{W}^{(n)} = \frac{W^{(n)}}{\rho W_b} = \frac{c}{2\rho W_b} \int_0^\infty |\tilde{E}^{(n)}|^2 r dr = \frac{B}{4} \int_0^\infty |\hat{E}^{(n)}(\hat{z}, \hat{r})|^2 \hat{r} d\hat{r} \left[\int_0^\infty \zeta S(\zeta) d\zeta\right]^{-1} ,$$

$$(4)$$

where $W_b = \mathcal{E}_0 I_0/e$ is the electron beam power. Thus, the exact solution of the initial-value problem for FEL amplifier with arbitrary gradient profile has been derived. As a result, if we have information on the eigenfunctions and eigenvalues, we are able to calculate radiation properties of the FEL amplifier which operates in high gain linear regime. The eigenvalue problem for the case of an arbitrary gradient axisymmetric profile is solved by means of the multilayer approximation method (see [13,14] for more details).

III   START-UP FROM SHOT NOISE

We start with the calculation of the average radiation power at the undulator exit. Under the accepted limitations we obtain that the contribution of the radiation with azimuthal index n to the total radiation power can be written in the following form [10,12]:

$$\langle\hat{W}^{(n)}\rangle = \frac{B}{\pi N_c} \int_{-\infty}^\infty d\hat{C} \left\{\sum_{k,j} u_k^{(n)}(u_j^{(n)})^* \exp\left\{[\lambda_k^{(n)} + (\lambda_j^{(n)})^*]\hat{z}\right\}\right.$$

$$\left. \times \int_0^\infty \Phi_{nk}(\hat{r})\Phi_{nj}^*(\hat{r})S(\hat{r})\hat{r}d\hat{r} \int_0^\infty \Phi_{nk}(\hat{r})\Phi_{nj}^*(\hat{r})\hat{r}d\hat{r}\right\} , \quad (5)$$

where $N_c = N_\lambda/(2\pi\rho)$, $N_\lambda = 2\pi I_0/(e\omega_0)$. The averaging symbol $\langle\cdots\rangle$ means the ensemble average over bunches.

At a sufficiently large undulator length the spectrum of the SASE radiation is concentrated within the narrow band near the resonance frequency ω_0. Therefore, the electric field of the wave can be presented as

$$E_x + iE_y = \tilde{E}(\vec{r}_\perp, z, t) e^{i\omega_0(z/c-t)} + \text{C.C.} ,$$

where $\tilde{E}$ is the slowly varying complex amplitude. Taking into account Parseval's theorem, and using the notation $\vec{\rho} = \vec{r}_\perp - \vec{r}_\perp'$ and $\vec{R} = (\vec{r}_\perp + \vec{r}_\perp')/2$, we can write the average angular spectrum in the form:

$$h(\vec{k}_\perp, z) = \frac{1}{(2\pi)^2} \int \gamma^{\text{eff}}(\vec{\rho}, z) \exp(-i\vec{k}_\perp \vec{\rho}) d\vec{\rho} \, ,$$

where we introduce the definition of effective transverse correlation function:

$$\gamma^{\text{eff}}(\vec{\rho}, z) = \frac{\int \langle \tilde{E}(\vec{R} + \vec{\rho}/2, z, t) \tilde{E}^*(\vec{R} - \vec{\rho}/2, z, t) \rangle d\vec{R}}{\int \langle |\tilde{E}(\vec{R}, z, t)|^2 \rangle d\vec{R}} \, .$$

Let us consider a axisymmetric electron beam. Using cylindrical coordinates we represent output power as a Fourier series in the azimuthal angle. In this case the angular spectrum can be written in the following dimensionless form:

$$h(\hat{\theta}, \hat{z}) = \left[\sum_{n,k,j} \int_{-\infty}^{\infty} d\hat{C} \left(\Omega_{kj}^{(n)} \int_0^{\infty} \Phi_{nk}(\hat{r}) J_n(\hat{\theta}\hat{r})\hat{r}d\hat{r} \int_0^{\infty} \Phi_{nj}^*(\hat{r}') J_n(\hat{\theta}\hat{r}')\hat{r}'d\hat{r}' \right) \right]$$

$$\times \left[2\pi \sum_{n,k,j} \int_{-\infty}^{\infty} d\hat{C} \left(\Omega_{kj}^{(n)} \int_0^{\infty} \Phi_{nk}(\hat{r})\Phi_{nj}^*(\hat{r})\hat{r}d\hat{r} \right) \right]^{-1} \, . \qquad (6)$$

To simplify this expression, we use the following notations: $\hat{\theta} = \theta \omega_0 r_0 / c$,

$$\Omega_{kj}^{(n)} = u_k^{(n)}(u_j^{(n)})^* \exp\left\{ [\lambda_k^{(n)} + (\lambda_j^{(n)})^*]\hat{z} \right\} \int_0^{\infty} \Phi_{nk}(\hat{r})\Phi_{nj}^*(\hat{r}) S(\hat{r})\hat{r}d\hat{r} \, .$$

IV   AN APPROACH FOR CONSTRUCTING TIME-DEPENDENT NUMERICAL SIMULATION CODES

Complete calculation of the parameters of the FEL amplifier can be performed only with numerical simulation codes. Time-dependent simulations of the FEL amplifier are being performed by simultaneous solutions of Maxwell's equations and the equations of motion of the electrons. To be specific, in this section we describe a three-dimensional, time-dependent code FAST developed for simulations of the SASE FEL [3].

The radiation field and the beam current density are presented in the form:

$$E_x + iE_y = \tilde{E}(\vec{r}_\perp, z, t) e^{i\omega_0(z/c - t)} + \text{C.C.} \, ,$$

$$j_z = -j_0(\vec{r}_\perp, z, t) + \tilde{j}_1(\vec{r}_\perp, z, t) e^{i\psi} + \text{C.C.} \, ,$$

where $\psi = k_w z + \omega_0(z/c - t)$, and $-j_0$ is the ensemble averaged current density.

We solve the electrodynamic problem using the paraxial approximation. In this case the wave equation may be written in the following form:

174

$$\left[\vec{\nabla}_{\perp} + 2\,i\frac{\omega_0}{c} \left(\frac{\partial}{\partial z} + \frac{1}{c}\frac{\partial}{\partial t} \right) \right] \tilde{E}(\vec{r}_{\perp}, z, t) = -\frac{4\pi\,i\theta_s\omega_0}{c^2}\tilde{j}_1(\vec{r}_{\perp}, z, t)\,. \tag{7}$$

The solution of this equation is

$$\tilde{E}(\vec{r}_{\perp}, z, t) = \frac{i\theta_s\omega_0}{c^2} \int\limits_0^z \frac{dz'}{z - z'} \int d\vec{r}_{\perp}'\,\tilde{j}_1\left(\vec{r}_{\perp}', z', t - \frac{z - z'}{c} \right) \exp\left[\frac{i\omega_0|\vec{r}_{\perp} - \vec{r}_{\perp}'|^2}{2c(z - z')} \right]\,. \tag{8}$$

The complex amplitudes, $\tilde{E}$ and $\tilde{j}_1$, are expanded in a Fourier series in the angle φ,

$$\tilde{E} = \sum_{n=-\infty}^{\infty} \tilde{E}^{(n)}(r, z, t)\,e^{-in\varphi}\,, \qquad \tilde{j}_1 = \sum_{n=-\infty}^{\infty} \tilde{j}_1^{(n)}(r, z, t)\,e^{-in\varphi}\,.$$

Then we get from (8) the expression for the Fourier harmonics:

$$\tilde{E}^{(n)}(r, z, t) = \frac{2\pi\theta_s\omega_0}{c^2}\,e^{-in\pi/2} \int\limits_0^z \frac{dz'}{z - z'} \int\limits_0^{\infty} dr'\,r'\,\tilde{j}_1^{(n)}\left(r', z', t - \frac{z - z'}{c} \right)$$
$$\times J_n\left(\frac{\omega_0 r r'}{c(z - z')} \right) \exp\left[\frac{i\omega_0(r^2 + r'^2)}{2c(z - z')} \right]\,. \tag{9}$$

Prior to the detailed analysis of start-up from noise (i.e. the self-consistent solution of (7) and the equations of particle motion under the shot noise initial conditions at the undulator entrance), it is relevant to discuss the region of applicability of the paraxial wave equation (7). The paraxial approximation assumes complex amplitude $\tilde{E}(\vec{r}_{\perp}, z, t)$ to be a slowly varying function on the scale of the radiation wavelength. When we consider start-up from noise, the beam current, $\tilde{j}_1(\vec{r}_{\perp}, z, t)$, is not a slowly varying function. The first limitation on the problem parameters means that the undulator should be sufficiently long, $k_w z \gg 1$. When the latter condition is fulfilled, we still cannot expect correct results in the three-dimensional case. Indeed, the incoherent undulator radiation has a wide continuous spectrum. When $k_w z \gg 1$, (7) correctly describes the fields in the narrow frequency band near the resonance frequency only, $\Delta\omega/\omega_0 \ll 1$. In terms of the far field zone, it gives correct results only for that part of the incoherent undulator radiation which is concentrated within the angle $\Delta\theta \ll 1/\gamma_z$ near the z axis.

When the FEL amplifier starts from the shot noise, a lot of transverse radiation modes are excited at the beginning of the amplification process; the radiation spectrum and the angular distribution in the far zone are relatively large. During the amplification process, the number of transverse radiation modes decreases, and the contribution of the coherent radiation into the total radiation power is increased. Also, the angular distribution of the radiation intensity in the far zone decreases. When it becomes much less than $1/\gamma_z$, we obtain a correct quantitative description of the amplification process starting from the shot noise.

175

One more relation, connecting the field and the current density, should come from the solution of the dynamical problem. When the space charge field can be neglected, the equations of motion may be written in the form:

$$\frac{\mathrm{dP}}{\mathrm{d}z} = -\frac{ie\theta_s}{2}\tilde{E}\,e^{i\psi} + \mathrm{C.C.}\,, \qquad \frac{\mathrm{d}\psi}{\mathrm{d}z} = \omega P/(c\gamma_z^2\mathcal{E}_0)\;.$$

To perform the simulations, we divide the electron beam into a large number of elementary volumes. The size of the divisions of the electron beam in the longitudinal direction should typically be chosen equal to the radiation wavelength. The number of azimuthal harmonics for calculations of the radiation field, N_φ, defines the number of azimuthal divisions of the electron beam. Typically, it should be by an order of magnitude larger than N_φ. Finally, the radial mesh should be chosen. The simulations are performed with a macroparticle method. The number of macroparticles in each volume is equal to N_m. At each integration step we calculate the bunching, $\hat{a}_1$, in each elementary volume:

$$\hat{a}_1 = \frac{1}{N_\mathrm{m}}\sum_{m=1}^{N_\mathrm{m}} e^{-i\psi_m}\;.$$

These values are used to calculate the azimuthal harmonics. The radiation field of the nth azimuthal harmonic in the discrete representation is calculated using the rigorous solution (9). At the next integration step, the sum of the azimuthal harmonics of the field is substituted into the equations of macroparticles motion in each volume, etc. As a result, one can trace the evolution of the radiation field and the particle distribution when the electron beam passes the undulator.

The initial shot noise in the electron beam is simulated according to the algorithm described in [15]. Since the actual number of particles per elementary volume, N_v, is large, the bunching in each box is the sum of a large number of random phasors with fixed amplitudes and uniformly distributed on $(0, 2\pi)$ phases. Using the central limit theorem, we can conclude that the phases of the bunching parameters are also distributed uniformly and the squared modulus of the amplitudes, $|\hat{a}_1|^2$, are distributed by the negative exponential distribution:

$$p(|\hat{a}_1|^2) = \frac{1}{\langle|\hat{a}_1|^2\rangle}\exp\left(-\frac{|\hat{a}_1|^2}{\langle|\hat{a}_1|^2\rangle}\right)\;, \tag{10}$$

where $\langle|\hat{a}_1|^2\rangle = 1/N_\mathrm{v}$. So, a negative exponential random generator with a mean value of $1/N_\mathrm{v}$ is used to extract the values of $|\hat{a}_1|^2$ for each volume. The phases of $\hat{a}_1$ are produced by a random number generator for the uniform distribution from 0 to 2π. These values are directly used as input parameters for the linear simulation code. In the nonlinear simulation code the macroparticles are distributed in such a way that the resulting bunching corresponds to the target value of $\hat{a}_1$ in each elementary volume.

The output of the code are the arrays for the field values in the Fresnel diffraction zone. Typical slice for the radiation pulse is presented in Fig. 1. Figure 2 presents

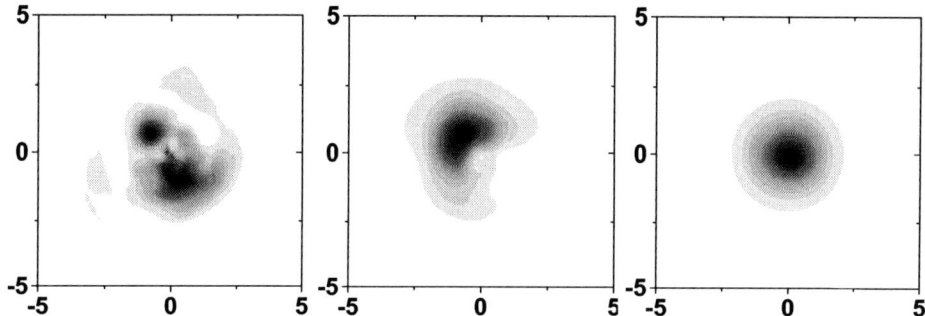

FIGURE 1. Distributions of the radiation intensity across one slice of the radiation pulse at different undulator lengths, $\hat{z} = 5$, $\hat{z} = 10$, and $\hat{z} = 15$ (left, middle, and right plots, respectively). The coordinates are normalized to $2^{1/2}\sigma_r$. Here $B = 1$, $\hat{\Lambda}_p^2 \to 0$, and $\hat{\Lambda}_T^2 = 0$. Calculations have been performed with linear simulation code FAST.

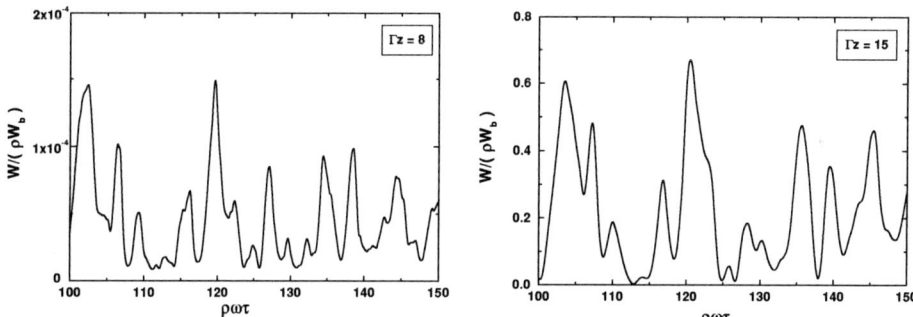

FIGURE 2. Temporal structure of the radiation pulse from the FEL amplifier starting from shot noise at the length of undulator of $\hat{z} = 8$ (left plot) and $\hat{z} = 15$ (right plot). Here $B = 1$, $\Lambda_p^2 \to 0$, $\Lambda_T^2 = 0$, and $\hat{z} = 15$. Calculations have been performed with linear simulation code FAST.

typical temporal structure of the radiation pulse from the FEL amplifier starting from shot noise. Post-processor programs are used to extract additional information for the field distribution in the far diffraction zone, for the spectrum, for the time, space and spectral correlation functions, and for the probability distributions of the radiation power and the radiation energy.

V   COMPARISON OF ANALYTICAL AND SIMULATION RESULTS

One can easily obtain that identical physical approximations have been used for analytical description of the high-gain linear regime and for numerical simulation algorithm. So, we should expect full agreement of the results in the high-gain linear regime. In other words, analytical results should serve as a primary standard for testing numerical simulation code. To be specific, we present the results for the following set of parameters: diffraction parameter $B = 1$, space charge parameter $\Lambda_p^2 \to 0$, energy spread parameter $\Lambda_T^2 = 0$, and $N_c = 7 \times 10^7$.

Analytical calculations have been performed taking into account nine beam radiation modes, TEM_{mn}, for $m, n = 0, 1, 2$. Numerical simulations have been performed with expansion of the radiation field up to the 6th azimuthal harmonic.

A Averaged radiation power

Left plot in Fig. 3 shows the evolution of the total radiation power from SASE FEL versus the undulator length. Simulation results have been obtained by means of averaging of the radiation power along the bunch (see Fig. 2). It is seen that analytical and simulation results agree well at $\hat{z} \gtrsim 7$.

Another interesting topic is partial contribution of different beam radiation modes into the total radiation power (see right plot in Fig 3). It is seen that both numerical and analytical results agree well at an increase of the undulator length. One can obtain that numerical simulations always give the value of the radiation power higher than analytical results. The reason is that the numerical simulation code calculates total gain, while the analytical formulae describe only the high-gain asymptote.

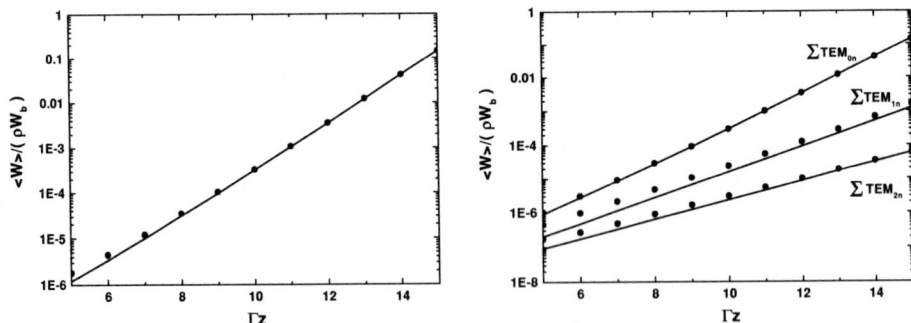

FIGURE 3. Averaged power (left plot) and partial contributions to the total power (right plot) versus undulator length for the FEL amplifier starting from shot noise. Here $B = 1$, $\Lambda_p^2 \to 0$, $\Lambda_T^2 = 0$, and $N_c = 7 \times 10^7$. Solid curves represent analytical results calculated with (5) for nine beam radiation modes ($m, n = 0, 1, 2$). The circles are the results obtained with linear simulation code FAST.

B   Averaged distributions of the radiation intensity in the near and far zone

Numerical simulation code produces an array containing values for the radiation field in the near zone (see Fig. 1). Averaging of these data over large number of slices gives us averaged distribution of the radiation intensity in the near zone. Relevant analytical quantity is given by expression (5). It is seen from the left plot in Fig. (4) that there is good agreement between simulation and analytical results.

Analytical predictions for the averaged angular distribution of the radiation power are given by (6). Similar characteristic can be also calculated with numerical simulation code. Using the values for the radiation field in the near zone, we find the radiation field propagating at any angle in the far diffraction zone. It is seen from right plot in Fig. 4 that both approaches agree well in the high-gain linear regime. It is important to stress that even after finishing the transverse mode selection process (which takes place after $\hat{z} \gtrsim 10$ for the considered numerical example) the distribution in the far zone differs visibly from the angular distribution of the fundamental TEM_{00} mode for maximum growth rate calculated in the framework of the steady-state theory (dotted line in Fig. 4).

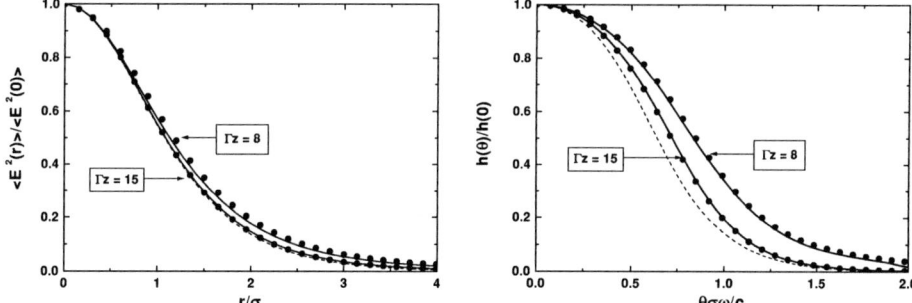

FIGURE 4. Averaged distributions of the radiation intensity in the near zone (left plot) and angular distribution of the radiation intensity in the far zone for the FEL amplifier starting from shot noise. Here $B = 1$, $\Lambda_p^2 \to 0$, $\Lambda_T^2 = 0$, and $N_c = 7 \times 10^7$. Solid curves represent analytical results calculated with (5) for sum of three radial modes ($n = 0, 1, 2$). The circles are the results obtained with linear simulation code FAST. Dashed lines represent distributions of the fundamental TEM_{00} mode for maximum growth rate calculated in the framework of the steady-state theory.

C   Averaged radiation spectrum

As we mentioned above, numerical simulation code produces an array containing values for the radiation field in the near zone. Integral spectrum of the radiation pulse can be calculated in the following way. Using the values for the radiation field in the near zone, we find the radiation field propagating at any angle in the far diffraction zone. At the next step of calculations we find the spectral distribution of the radiation power for each angle, and after integrating over all angles we obtain integral spectrum of the radiation pulse. Typical spectrum of the radiation pulse obtained from numerical simulations is present in Fig. 5.

Analytical results give predictions for the averaged radiation spectrum (see (5)). To obtain averaged spectrum from numerical simulation code, we performed large number of statistically independent simulation runs. Each run gives spectrum with spiky structure as shown in the left plot in Fig. 5. Average of a large number of radiation spectra is presented in the right plot in Fig. 5. It is seen that the analytical and simulation results for the averaged radiation spectrum agree well in the high-gain linear regime.

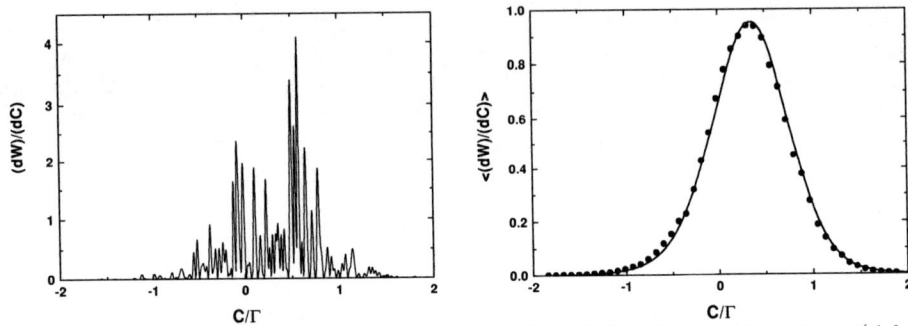

FIGURE 5. Typical spectrum of the radiation pulse (left plot) and averaged spectrum (right plot) of the radiation from the FEL amplifier starting from shot noise at the undulator length $\hat{z} = 15$. Here $B = 1$, $\Lambda_p^2 \to 0$, $\Lambda_T^2 = 0$, and $N_c = 7 \times 10^7$. Solid curve in the right plot represents analytical results calculated with (5) for nine beam radiation modes $(m, n = 0, 1, 2)$. The circles are the results obtained with linear simulation code FAST.

VI TRANSVERSE COHERENCE

In the case of axisymmetric electron beam the radiation field statistically isotropic. For such a field the effective correlation function depends only on the modulus $|\vec{\rho}|$ and the angular spectrum depends on the modulus $|\vec{k}_\perp|$. Thus, we have

$$\gamma_1^{\text{eff}}(\hat{\rho}, \hat{z}) = 2\pi \int_0^\infty J_0(\hat{\rho}\hat{\theta})h(\hat{\theta})\hat{\theta}d\hat{\theta} , \qquad (11)$$

where $\hat{\rho} = |\vec{\rho}|/r_0$. It is natural to define the area of coherence for an isotropic inhomogeneous field as:

$$S_{\text{coh}}(\hat{z}) = 2\pi \int_0^\infty |\gamma_1^{\text{eff}}(\hat{\rho}, \hat{z})|^2 \hat{\rho}d\hat{\rho} . \qquad (12)$$

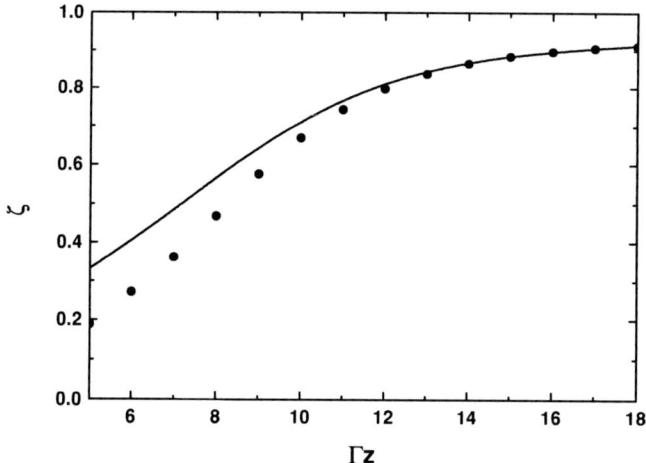

FIGURE 6. Degree of transverse coherence of the radiation from the FEL amplifier versus the undulator length. Solid curve represents analytical results, and the circles are the results obtained with linear simulation code FAST. Here $B = 1$, $\Lambda_p^2 \to 0$, $\Lambda_T^2 = 0$, and $N_c = 7 \times 10^7$.

Using the relation $S_{coh} = \pi \hat{r}_{coh}^2$, it is conventional to introduce the notion of the radius of coherence $\hat{r}_{coh}$.

To describe the formation of the transverse coherence, we should define the degree of coherence. One possible definition can be made as follows. After statistical analysis of the numerical results we find $\hat{r}_{coh}$. Then we find the radius of coherence $\hat{r}_{max}$ for the fully coherent radiation which is represented by the fundamental $\Phi_{00}(\hat{r})$ mode for maximum growth rate. The field distribution of this mode for Gaussian density distribution in the electron beam can be found by the multilayer approximation method described in section 2.2. The degree of coherence, ζ, may be defined as

$$\zeta = \hat{r}_{coh}^2 / \hat{r}_{max}^2 . \tag{13}$$

Using angular distributions of the radiation field in the far diffraction zone we can trace the dependence of the degree of transverse coherence versus undulator length. Solid line in Fig. 6 is the results of analytical calculations, and the circles are the results obtained with numerical simulation code. We can state that there is good agreement between analytical and simulation results. It is clearly seen that the degree of coherence differs visibly from the unity in the high-gain linear regime, $\zeta \simeq 0.9$ at $\hat{z} = 15$.

Another possible way to define the degree of coherence is based on the statistical analysis of fluctuations of the instantaneous power. Since in the linear regime we deal with a Gaussian random process, the power density at fixed point in space fluctuates in accordance with the negative exponential distribution [14]. If there is

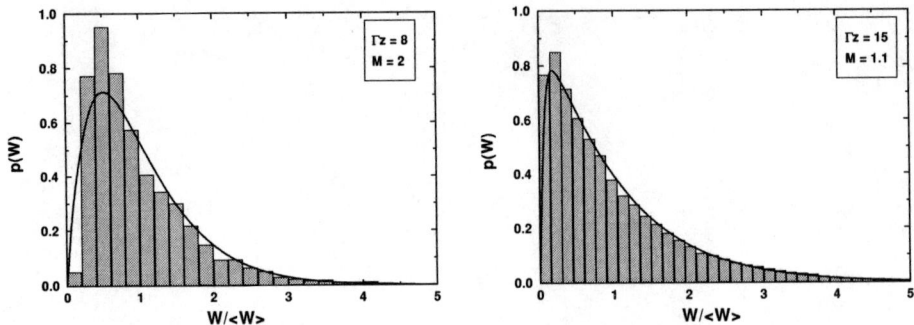

FIGURE 7. Histogram of the probability distribution of the instantaneous radiation power from the FEL amplifier starting from shot noise. at the length of undulator of $\hat{z} = 8$ (left plot) and $\hat{z} = 15$ (right plot). Here $B = 1$, $\Lambda_p^2 \to 0$, $\Lambda_T^2 = 0$, and $\hat{z} = 8$. Calculations have been performed with linear simulation code FAST. Solid lines represent gamma distribution with $M = 2$ (left plot) an $M = 1.1$ (right plot).

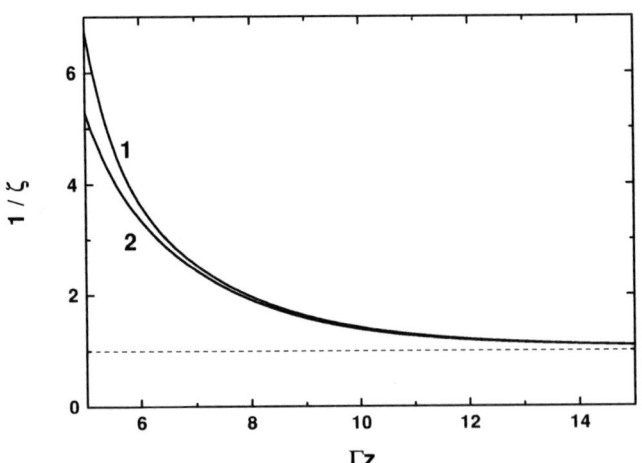

FIGURE 8. Inverse value of transverse coherence versus undulator length. Here $B = 1$, $\Lambda_p^2 \to 0$, $\Lambda_T^2 = 0$, and $N_c = 7 \times 10^7$. Calculations have been performed with linear simulation code FAST. Curve 1 is calculated using instantaneous fluctuations of the radiation power. Curve 2 is calculated using angular distribution of the radiation power in far zone.

full transverse coherence then the same refers to the instantaneous power W equal to the power density integrated over cross section of the radiation pulse. If the radiation is partially coherent, then we have a more general law for instantaneous power fluctuations, namely the gamma distribution [14,16]:

$$p(W) = \frac{M^M}{\Gamma(M)} \left(\frac{W}{\langle W \rangle}\right)^{M-1} \frac{1}{\langle W \rangle} \exp\left(-M\frac{W}{\langle W \rangle}\right) , \tag{14}$$

where $\Gamma(M)$ is the gamma function with argument M, and

$$M = \frac{1}{\sigma_W^2} , \qquad \sigma_W^2 = \langle (W - \langle W \rangle)^2 \rangle / \langle W \rangle^2 , \tag{15}$$

The parameter $M = 1/\sigma_w^2$ of this distribution can be considered as the number of transverse modes. Then the degree of coherence in the linear regime, ζ, may be defined as follows

$$\zeta = \frac{1}{M} = \sigma_w^2 . \tag{16}$$

The value of M should be calculated with numerical simulation code producing time-dependent results for the radiation power (see Fig. 2). Figure 7 presents the probability distribution of the fluctuations of the radiation power for two different undulator lengths of $\hat{z} = 8$ and $\hat{z} = 15$. In Fig. 8 we present the dependence of the number of transverse modes on the undulator length for the specific value $B = 1$ of the diffraction parameter. It is seen that both definitions for the degree for the transverse coherence, (13) and (16), are consistent in the high-gain linear regime.

VII   DISCUSSION

Let us discuss asymptotical behaviour of the degree of transverse coherence. At a large value of the undulator length it approaches to unity asymptotically as $(1 - \zeta) \propto 1/z$, but not exponentially, as one can expect from simple physical assumption that transverse coherence establishes due to the transverse mode selection. The latter effect indeed takes place as it is illustrated in Fig. 1. That is why the degree of coherence grows quickly at an early stage of amplification. Starting from some undulator length the contribution to the total power of the fundamental mode becomes to be dominant (see Fig. 3). However, one should take into account that the spectrum width has always finite value (see Fig. 5). Actually this means that in the high gain linear regime the radiation of the SASE FEL is formed by many fundamental TEM$_{00}$ modes with different frequencies. The transverse distribution of the radiation field of the mode is also different for different frequencies. As a result of interference of these modes we do not have full transverse coherence. Taking into account this consideration, we can simply explain asymptotical behaviour of the degree of transverse coherence – this is reflection of the slow evolution of the width of the radiation spectrum as $z^{-1/2}$ with the undulator length.

All the results presented above have been obtained in the framework of the linear theory. Simulations with nonlinear code shows that for the considered numerical example the saturation occurs at $\hat{z} \simeq 18$. Using the plot presented in Fig. 6 we find that the value of the transverse coherence is less than 0.9 in the end of the linear regime. A typical range of the values of N_c is 10^6-10^9 for the SASE FEL of wavelength range from X-ray up to infrared. The numerical example presented in this paper is calculated for $N_c = 7 \times 10^7$ which is typical for a VUV FEL. It is worth to mention that the dependence of the saturation length of the SASE FEL on the value of N_c is rather weak, in fact logarithmic (see (5)). Therefore, we can state that obtained effect limiting the value of transverse coherence might be important for practical SASE FELs.

REFERENCES

1. Fawley, W.M., *An Informal Manual for GINGER and its post-processor XPLOT-GIN*, **LBID-2141, CBT Tech Note-104, UC-414** *(1995)*.
2. Reiche, S., *Nucl. Instr. and Methods* **A429**, *243 (1999)*.
3. Saldin, E.L., Schneidmiller, E.A., and Yurkov, M.V., *Nucl. Instr. and Methods* **A429**, *233 (1999)*.
4. *A VUV Free Electron Laser at the TESLA Test Facility: Conceptual Design Report*, **DESY Print TESLA-FEL 95-03**, *Hamburg (1995)*.
5. Rossbach, J., *Nucl. Instr. and Methods bf A375, 269 (1996)*.
6. *Conceptual Design of 500 GeV e^+e^- Linear Collider with Integrated X-ray Facility (Editors Brinkmann, R. et al.)* **DESY 1997-048, ECFA 1997-182**, *Hamburg (1997)*.
7. *Linac Coherent Light Source (LCLS) Design Study Report (The LCLS Design Study Group, Stanford Linear Accelerator Center)* **SLAC-R-521** *(1998)*.
8. Moore, G.T., *Opt. Commun.*, **52**, *46 (1984)*.
9. Moore, G.T., *Nucl. Instr. and Methods bf A250, 381 (1986)*.
10. Kim, K.J, *Phys. Rev. Lett.*, **57**, *1871 (1986)*.
11. van Kampen, N.G., *Physica*, **21**, *949 (1951)*.
12. Krinsky,S., and Yu, L.H., *Phys. Rev.*, **A35**, *3406 (1987)*.
13. Saldin, E.L., Schneidmiller, E.A., and Yurkov, M.V., *Opt. Commun.*, **97**, *272 (1993)*.
14. Saldin, E.L., Schneidmiller, E.A., and Yurkov, M.V., *The Physics of Free Electron Lasers, Springer, Berlin (1999)*.
15. Saldin, E.L., Schneidmiller, E.A., and Yurkov, M.V., *Opt. Commun.*, **148**, *383 (1998)*.
16. Saldin, E.L., Schneidmiller, E.A., and Yurkov, M.V., *Nucl. Instr. and Methods* **A429**, *229 (1999)*.

Present Status of X-Ray FELs[1]

Kwang-Je Kim and Zhirong Huang

Advanced Photon Source, Argonne National Laboratory, Argonne, IL 60439, USA

Abstract. We review the current status of the theoretical and experimental efforts to develop high-gain free-electron lasers for generating x-ray beams. Such a source will provide ten orders of magnitude brightness enhancement and two orders of magnitude time resolution compared to the current third-generation sources.

I THE PROMISE

The third-generation synchrotron radiation facilities have been highly successful in machine operation and in scientific output. The success derives from the high brightness of the spontaneous emission from undulators in the straight sections of optimized electron storage rings. The brightness (in units of photons per sec per mm^2 per mrad^2 per 0.1% relative bandwidth) is about 10^{20} in the wide spectral range from visible to hard x-ray wavelengths. Additional characteristics are the tunability of spectrum and the adjustability of the polarization. The photon beams consist of bunches that are about a few tens of picoseconds long. The bunch length defines the minimum time resolution. The peak brightness during the bunch duration is about 10^{23}.

For the spontaneous radiation discussed above, the radiation intensity is an incoherent sum of contributions from individual electrons. For a high-brightness electron beam passing through a long undulator, the radiation-electron beam system becomes a high-gain free-electron laser (FEL) amplifier [1,2]. If coherent input radiation were available, the high-gain amplifier would be an intense, coherent source. For x-rays, the amplifier configuration is not practical due to the absence of the coherent input source. However, even without the input coherent radiation, intense, quasi-coherent radiation can be produced from amplification of the initial spontaneous radiation when the gain of the system is extremely high. Such a configuration is referred to as self-amplified spontaneous emission (SASE).

SASE in the x-ray region appears to be realizable in view of the recent advance in the production of bright electron beams by means of a photocathode rf gun, bunch

---

[1] Work supported by the U. S. Department of Energy, Office of Basic Energy Sciences, under Contract No. W-31-109-ENG-38.

CP581, *Physics of, and Science with, the X-Ray Free-Electron Laser,* edited by S. Chattopadhyay et al.
2001 American Institute of Physics 0-7354-0022-9

185

TABLE 1. Peak brightness enhancement from undulator radiation to SASE.

	Undulator	SASE	Enhancement factor
Number of photons	$N_e/137$	$N_{lc}N_e/137$	$N_{lc} \sim 10^6$
Transverse phase space	$(2\pi\epsilon_x)(2\pi\epsilon_y)$	$(\lambda/2)^2$	10^2
Pulse duration	10 ps	100 fs	10^2
Peak brightness	10^{23}	10^{33}	10^{10}

compressors, and a high-energy linear accelerator (linac). The Linac Coherent Light Source (LCLS) is a proposal to use the SLAC linac to produce and use SASE at one Angstrom [3]. The SASE beam in this proposal is about one million times more intense than spontaneous undulator radiation. It is coherent transversely (or diffraction limited), with the transverse phase-space area about one hundred times smaller than the partially coherent undulator radiation. The bunch length, being squeezed by the bunch compressor in the begining of the linac for higher peak current, is about 100 femtoseconds, or two orders of magnitude shorter than that of the third-generation sources. The peak brightness of the x-ray SASE is projected to be about ten orders of magnitude higher than that available from third-generation synchrotron radiation sources, i.e., 10^{33} (in normal units). Table 1 summarizes the brightness enhancement factors. The phenomenal increase in peak brightness and the subpicosecond time resolution would open up exciting scientific applications [4].

To fully utilize the capability of SASE, a "fourth generation" user facility [5] will need to employ a superconducting linear accelerator that is operable at a high repetition rate and can drive a farm of SASE and undulator devices, as shown in Fig. 1. The TESLA FEL proposal is a concrete example of a fourth-generation facility [6].

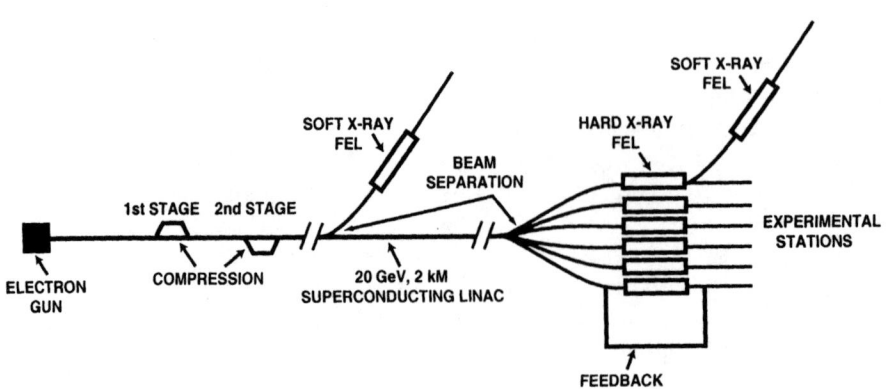

FIGURE 1. Fourth-generation x-ray facility using SASE FELs (from Ref. 5).

II   DEVICE REQUIREMENTS

The general components of a linac-based SASE generation are: an rf photo-cathode gun with emittance corrector, a set of bunch compressors, an acceleration section, and a long undulator.

All accelerators used in experiments to demonstrate SASE take advantage of the low emittance that can be realized with the rf photocathode guns invented at Los Alamos National Laboratory [7] and the emittance compensation scheme [8]. The typical parameters of electron bunches from present rf photocathode guns are: normalized rms emittance $\gamma\epsilon_{x,y}$ of about 5 mm-mrad, bunch charge about 0.5 nC, and pulse length of a few ps. These are adequate for SASE experiments for wavelengths 100 nm or longer. The performance required for a 1-Å SASE, 1 mm-mrad emittance for a 1-nC bunch, appears to be feasible by enhancing various components [9]: optimizing the accelerating gradient and the rf phase, shaping the transverse and longitudinal profiles of the laser pulse, etc.

Although the electron beam from an rf photocathode gun is already bunched to a few picoseconds, it needs to be further compressed to increase the peak current to a few kA to drive an x-ray FEL. In addition to the space charge and the wake-field effect, the bunch compressors must be designed carefully [10] to minimize the emittance dilution effect due to the coherent synchrotron radiation (CSR) [11].

The emittance could also be diluted in the course of the acceleration process in linacs, not only from single particle effects such as focusing mismatch, dispersive effects, and chromatic effects, but also from collective effects due to impedance. Some of these effects, which are due to correlations between different degrees of freedom, can be corrected [10]. Emittance dilution has been extensively studied in connection with linear collider design efforts [12]. Experimentally, it was shown that the increase in the normalized vertical emittance in the SLAC linac, which is about 1.5 mm-mrad in the beginning, can be controlled to less than 50% through the 3-km, 50-GeV acceleration. This is about the level of the control required for the LCLS.

The x-ray SASE requires a long undulator, typically about 100 m long, containing about 1000 periods. The undulator can be divided into several segments with the space between the segments used for diagnostics, focusing, and pumping. The segment must be aligned to an accuracy of a few μm. The tolerance on the magnetic field in each segment is also tight, but within the current state of the art, due to recent developments in undulator construction spurred by the needs of the third-generation synchrotron radiation facilities [13].

III   THEORETICAL DEVELOPMENT

The basic theories of high-gain FELs have been well developed in the linear regime [14–17]. Taking into account the effects of beam energy spread, emittance, radiation diffraction, and optical guiding, the field amplitude at frequency ω can be written as

$$E_\omega(\boldsymbol{x}; z) = \sum_n C_n E_n(\boldsymbol{x}) e^{-i\mu_n 2\rho k_u z} + \text{continuum modes.} \tag{1}$$

Here $\boldsymbol{x}$ is the transverse coordinate, z is the distance into the undulator, $k_u = 2\pi/\lambda_u$, λ_u is the undulator period, μ_n and E_n are the discrete and complex solutions for the eigenvalue and the eigenmode of the Maxwell-Klimontovich equations in the frequency domain [18–20], respectively, and C_n is the mode expansion coefficient found by solving the FEL initial value problem [21,22]. The dimensionless parameter ρ is important in determining basic scaling properties of a high-gain FEL [2], and is given by

$$\rho = \left(\frac{eK^2[JJ]^2 Z_0 j}{32\gamma_0^3 mc^2 k_u^2} \right)^{1/3}, \tag{2}$$

where K is the undulator parameter, $[JJ]$ is the Bessel factor associated with the planar undulator, $Z_0 = 377\ \Omega$, j is the electron current density, and $\gamma_0 mc^2$ is the beam energy. Typically ρ is about 5×10^{-4} for the planned x-ray SASE projects.

A   Start-up, Exponential Growth, and Saturation

In the linear regime before saturation, the FEL power spectrum can be expressed as [22]

$$\frac{dP}{d\omega} = g_A \left[\left(\frac{dP}{d\omega} \right)_0 + g_S \frac{\rho\gamma_0 mc^2}{2\pi} \right] \exp \left[\frac{z}{L_G} - \frac{(\Delta\omega)^2}{2\sigma_\omega^2} \right], \tag{3}$$

where $(dP/d\omega)_0$ is the external coherent power spectrum, $L_G = \chi\lambda_u/(4\pi\sqrt{3}\rho)$ is the power gain length, $\chi \geq 1$ is conveniently related to the beam parameters by the interpolating formula given in Ref. [23], and σ_ω is the gain bandwidth. The numerical factors g_A and g_S determine the input coupling to the exponentially growing mode and the effective noise power in units of $\rho\gamma_0 mc^2/(2\pi)$, respectively. It is found that g_A increases from 1/9 to 1 with the increasing initial energy spread for a Gaussian energy distribution [24]. The white noise spectrum, $g_S\rho\gamma_0 mc^2/(2\pi)$, can be interpreted as the coherent fraction of the spontaneous undulator radiation in the first two power gain lengths [25,22]. Thus, g_S increases (from 1) with energy spread and emittance through the increase in gain length. For x-ray SASE FELs, $G = g_A g_S$ can be significantly larger than the one-dimensional, cold beam value (1/9) [21].

As the radiation builds up exponentially after the first two power gain lengths, the phase of most electrons relative to the radiation field changes continuously and eventually the electron beams start to gain energy. At $z = z_{\text{sat}} \approx \lambda_u/\rho \approx 20 L_G$ the

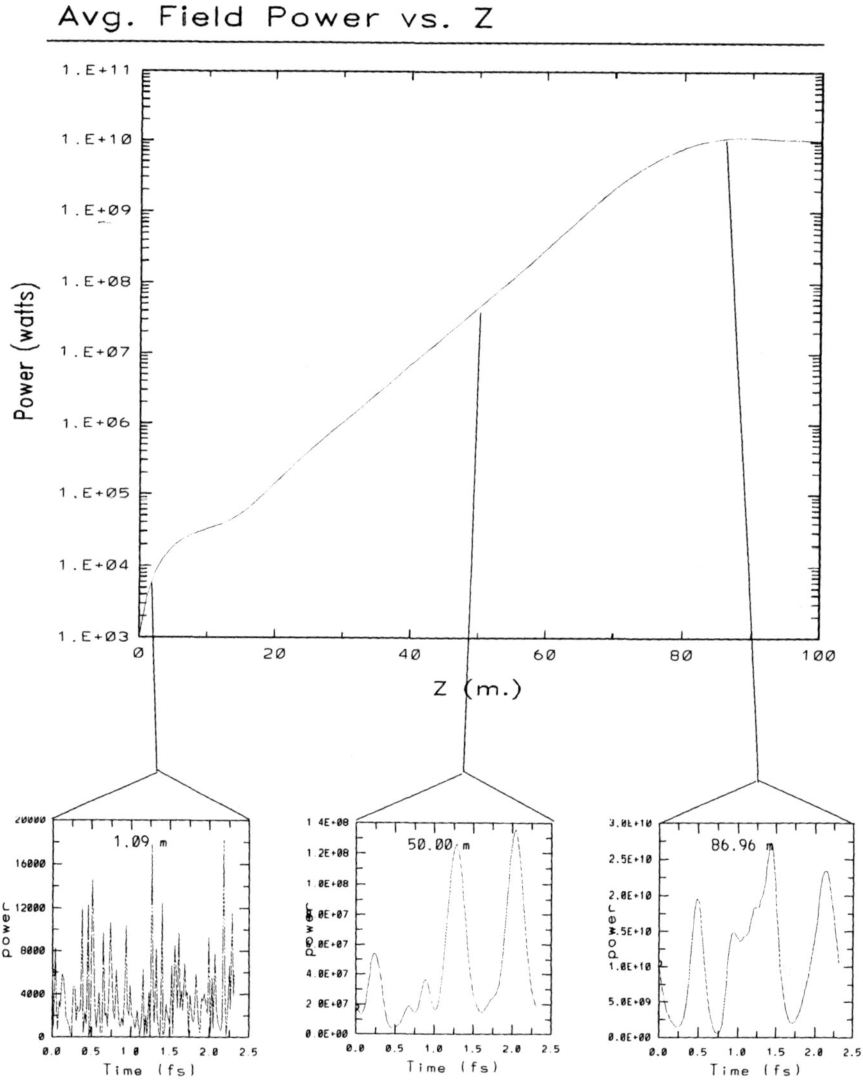

FIGURE 2. Evolution of the LCLS radiation power and the temporal coherence (courtesy of H.-D. Nuhn, SLAC).

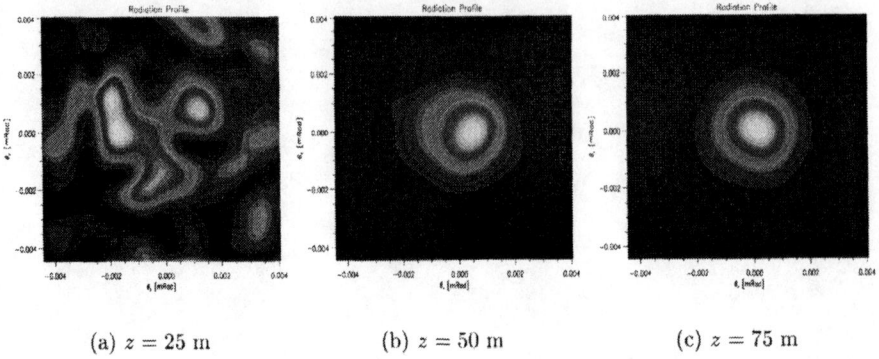

(a) $z = 25$ m   (b) $z = 50$ m   (c) $z = 75$ m

FIGURE 3. Evolution of the LCLS transverse profiles at different z location (courtesy of S. Reiche, UCLA).

growth of radiation power stops, i.e., *saturates*. A typical power evolution for the LCLS is obtained by GINGER simulation and is shown in Fig. 2. Saturation effects can be studied by a quasilinear extension of the linear theory [16]. The saturated power obtained by this quasilinear approach [26] agrees with an empirical formula obtained through numerical simulations [23], i.e.,

$$P_{\text{sat}} \approx \frac{1.6}{\chi^2} \rho P_{\text{beam}},$$ (4)

where χ is the gain length degradation factor and P_{beam} is the total electron beam power. After saturation, electrons trapped in the ponderomotive potential of the combined undulator and radiation fields execute synchrotron oscillations, giving rise to the development of sideband frequencies and amplitude modulation of the radiation field [27].

B   Transverse Mode Evolution and Coherence

A remarkable feature in the exponential growth regime is optical guiding, the phenomena in which the transverse profile of the radiation beam is frozen. This arises because the field amplitude is dominated by the fundamental mode with the largest growth rate, the imaginary part of μ_1. Higher-order modes usually have a much smaller growth rate and hence are negligible after a few e-folding of the fundamental modes [28]. This process is illustrated in Fig. 3's snapshots of radiation angular patterns at different z locations.

For an axisymmetric electron beam, the fundamental mode of the radiation can be approximated by

$$E_1(\boldsymbol{x}) = \exp\left[-\alpha(\omega)\frac{r^2}{\sigma_x^2}\right],$$ (5)

190

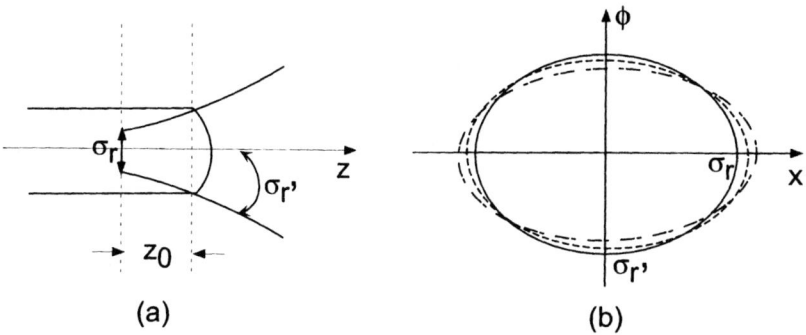

FIGURE 4. Illustration of FEL fundamental mode and the smearing of the transverse coherence.

where $\alpha = \alpha_R + i\alpha_I$ is a complex mode parameter, $r = \sqrt{x^2 + y^2}$, and σ_x is the rms transverse size of the electrons. Such a mode (Fig. 4 (a)) has its minimum waist located

$$z_0 = -\frac{\pi \sigma_x^2}{\lambda} \frac{\alpha_I}{\alpha_R^2 + \alpha_I^2} \tag{6}$$

away from the undulator exit with a Rayleigh length

$$L_R = \frac{\pi \sigma_x^2}{\lambda} \frac{\alpha_R}{\alpha_R^2 + \alpha_I^2}. \tag{7}$$

The rms waist size and the far-field divergence angle are

$$\sigma_r = \sqrt{\frac{\lambda L_R}{4\pi}}, \quad \sigma_{r'} = \sqrt{\frac{\lambda}{4\pi L_R}}. \tag{8}$$

The dominance of a single Gaussian mode means that the SASE FEL at a given frequency is transversely coherent, even though the emittance of the electron beam could be larger than the diffraction limit $\lambda/(4\pi)$. However, taking into account the finite SASE bandwidth, the overall radiation is formed by many fundamental modes with slightly different frequencies and mode parameters; hence the transverse coherence of the overall SASE radiation is somewhat degraded [29]. This degrading effect is illustrated in Fig. 4 (b) as the smearing of the transverse phase ellipses of the radiation beam. Each ellipse, representing a distinct fundamental mode at a particular frequency ω, has almost the same area $\lambda/4$ since the spread $\Delta\lambda/\lambda = \Delta\omega/\omega \sim \rho \ll 1$. However, ellipses are not properly aligned to each other due to the frequency dependence of α in z_0 and L_R. The resulting emittance of the radiation is larger than the minimum emittance $\epsilon_0 = \lambda/(4\pi)$. To quantify the degree of transverse coherence, we introduce a transverse mode number

$$M_T = \left(\frac{\epsilon_r}{\epsilon_0}\right)^2 \tag{9}$$

with the rms emittance of the radiation beam ϵ_r given by

$$\epsilon_r = \sqrt{\langle x^2 \rangle \langle \phi^2 \rangle - \langle x\phi \rangle^2}. \tag{10}$$

Full transverse coherence means that $M_T = 1$. For the LCLS, we find that $M_T \approx 1.06$ before saturation [30], indicating a high degree of transverse coherence for the proposed SASE FEL.

C Temporal Coherence and Intensity Fluctuation

In the exponential growth regime, the spectral bandwidth σ_ω of the SASE radiation decreases as a function of the distance into the undulator as [14,15]

$$\sigma_\omega \approx \omega_0 \sqrt{\frac{\rho}{z/\lambda_u}}, \quad z \leq z_{\text{sat}}. \tag{11}$$

At the saturation point $z_{\text{sat}} \approx \lambda_u/\rho$ the relative bandwidth becomes $\sigma_\omega/\omega_0 \sim \rho$. Beyond this point, the bandwidth starts to increase due to the development of synchrotron sidebands.

In the time domain the SASE radiation consists of N_e (total number of electrons) wavepackets, each with an rms duration

$$\sigma_\tau = \frac{1}{2\sigma_\omega} \approx \frac{1}{2\omega_0} \sqrt{\frac{z/\lambda_u}{\rho}}. \tag{12}$$

The time domain picture was emphasized by Bonifacio *et al.* [31]. The wavepackets are contained within the bunch length cT of the electron beam, exhibiting M_L coherent regions (spikes). Here the longitudinal mode number

$$M_L \approx \frac{T}{\sigma_\tau} = \frac{T\sigma_\omega}{2} \tag{13}$$

decreases as z increases before saturation (see the inserts of Fig. 2). In general, the pulse length of the radiation field is shorter than that of the electron beam with a smooth current profile because the gain is strongest when the longitudinal current of the electrons is highest. Thus, the number of wavepackets is smaller than given by Eq. (13).

Statistical properties of SASE light are determined by the fact that the field amplitude E_ω is proportional to the sum of a large number of random phase factors. Light with these properties, such as sunlight or spontaneous undulator radiation, is referred to as "chaotic." This topic has been extensively discussed, for example by Goodman [32]. In the context of SASE, it is thoroughly discussed in Ref. [33]. A simple review can be found in Ref. [34]. The probability distribution of the field amplitude E_ω of a chaotic light is Gaussian, as a straightforward application of the central limit theorem. Equivalently, the intensity at a given frequency $I_\omega = |E_\omega|^2$ has an exponential probability distribution in which the variance is equal to the average intensity. The fluctuation is therefore 100%. In general, we consider a partial flux ΔW as the flux within a 6-D phase-space volume $\Delta \Omega$. The probability distribution of ΔW is given by the "gamma" probability distribution [32]. In the gamma distribution, the fluctuation is reduced by a factor $\sqrt{M}$, where M is the number of coherent modes in $\Delta \Omega$. We can write $M = M_T M_L$, where M_T given in Eq. (9) and M_L given in Eq. (13) are the transverse and longitudinal mode numbers, respectively.

We can now compare the fluctuation in SASE from the LCLS and in undulator radiation from typical third-generation light sources at x-ray wavelengths. For the SASE at saturation, $M_T \approx 1$ (almost full transverse coherence), $\sigma_\omega \approx \rho\omega$, $\rho \approx 10^{-3}$, T is about 100 fs, while for the latter $M_T \approx \epsilon_x \epsilon_y / \epsilon_0^2 \gg 1$, $\sigma_\omega \approx 0.01\omega$, and T is about 100 ps. Therefore the fluctuation in SASE is larger by at least two orders of magnitude than that in undulator radiation.

For most of the SASE demonstration experiments discussed in Section V, the electron bunch lengths with respect to the coherence length are relatively short so that the fundamental SASE fluctuations are large and readily measurable. The fluctuation measurement is therefore a useful diagnostic tool in such cases. For the case of the planned x-ray FEL, however, the overall fluctuation is dominated by the fluctuations from accelerator jitter because the fluctuation from the fundamental SASE property is relatively small.

D   Induced Harmonics

When an electron beam is strongly bunched in the ponderomotive potential formed by the undulator magnetic field and the fundamental radiation field, substantial higher-harmonic bunching is induced. Thus, FELs that employ planar undulators are capable of generating intense, coherent harmonic radiation, especially at odd harmonic frequencies.

These induced harmonics have been studied in the context of high-gain FELs using a 1-D model [35] and the 3-D simulation code MEDUSA [36]. Recently, a 3-D theory of harmonic generation has been developed [37], using the coupled Maxwell-Klimontovich equations, that includes electron energy spread and emittance, the radiation diffraction and optical guiding. In general, each harmonic field is a sum of a linear amplification term and a term driven by nonlinear harmonic interactions. After a certain stage of exponential growth, the dominant nonlinear term is determined by interactions of the lower nonlinear harmonics and the fundamental radiation. As a result, the gain length, transverse profile, and temporal structure of the first few harmonics are eventually governed by those of the fundamental. For example, driven by the third power of the radiation field in the fundamental, the third nonlinear harmonic grows three times faster, has an equally coherent transverse mode (with a smaller spot size), and has a more spiky temporal struc-

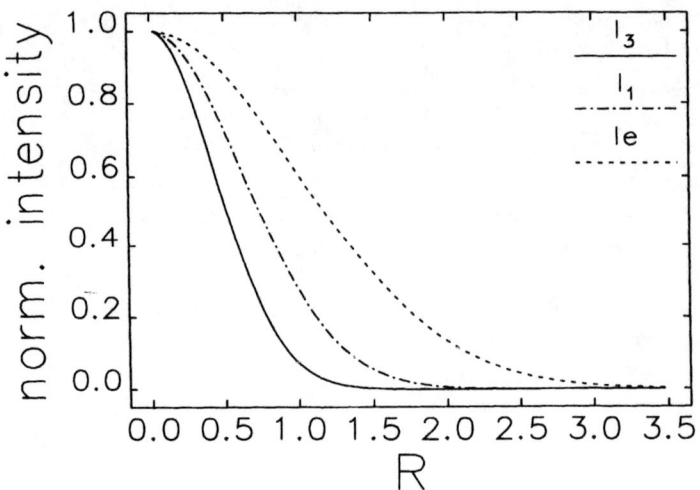

FIGURE 5. Transverse profiles of the LCLS third harmonic (I_3), the fundamental radiation (I_1), and the electron beam (I_e) as functions of the radius R in units of electron beam size.

ture than the fundamental of SASE FELs. For LCLS, the transverse profiles of the third nonlinear harmonic and the fundamental radiation are calculated numerically in the exponential growth regime (see Fig. 5). The evolution of the third-harmonic power is given by [37]

$$\frac{P_3}{\rho P_{\text{beam}}} = 0.11 \left(\frac{P_1}{\rho P_{\text{beam}}} \right)^3 . \tag{14}$$

Taking $P_1 = P_{\text{sat}}/2 \approx 4$ GW before the FEL saturation, one can estimate the third-harmonic power to be 15 MW, about 0.4% of the fundamental power level.

The nonlinear harmonic generation occurs naturally in one long undulator for a SASE FEL and a seeded FEL. As long as the laser fundamental saturates after a certain length of undulator, the induced harmonics are generated around certain levels, much less sensitive to the e-beam parameters, undulator errors, and wake-field effects than other (linear) harmonic generation schemes. The most significant nonlinear harmonic generation is the third-harmonic radiation, typically around one percent of the fundamental power level near saturation. Even harmonics are also present due to the transverse gradient of the beam current. They normally have much lower power levels than their odd counterparts [36,37]. Finally, such a harmonic generation mechanism may be utilized to reach shorter radiation wave-lengths or to relax some stringent requirements on the electron beam quality for x-ray FELs.

E   Quantum Effects

There are several quantum corrections to the SASE properties. However, these effects are all negligible in the case of x-ray SASE as discussed below [38,39].

First, the quantum correction to the classical gain formula is small if the recoil energy is small compared to the gain bandwidth, or the photon energy is smaller than the electron energy spread. This condition is well satisfied for the x-ray SASE parameters.

Second, the effective noise signal needs to be modified when more than one electron occupies the quantum mechanical unit cell of volume λ_c^3, where λ_c is the electron's Compton wavelength. This is far from the case in the x-ray SASE.

Third, after taking the quantum effect into account, the mode number M becomes $M/(1 + 1/\delta)$, where δ, known as the degeneracy number, is the number of photons per mode. This correction is also negligible in the x-ray SASE, since $\delta \gg 1$.

F   Simulation Codes

An important tool in designing and analyzing FEL performance is the use of simulation codes with macroparticle models. Each particle, representing tens of thousands of electrons, follows the equations of motion in the combined undulator and radiation fields. The radiation field generated by thousands of simulation particles is then found by solving the paraxial wave equation. A extensive review of FEL simulation codes is given in Ref. [40]. The cross comparison between codes and with theory are generally in good agreement [41]. A steady state code such as TDA [42] is seeded by a monochromatic external radiation field, and hence no variation is allowed along the electron bunch. This approximation permits simulating the electron beam with one slice that is equal to one radiation wavelength λ, thus reducing the requirements for computer resources. MEDUSA, another steady-state code, is extended to include higher hamonics as well as sidebands beside the fundamental frequency [36]. In "time-dependent" simulation codes such as GINGER [43] and GENESIS [44] (both can be run in the steady-state mode), many electron and photon slices along the bunch longitudinal coordinates are tracked by applying an appropriate slippage condition. The time-dependent simulation allows variations along the bunch and is therefore polychromatic around the resonant frequency. Starting up from shot noise can be handled by loading an appropriate amount of random deviation from the uniform longitudinal distribution of the simulating particles [45]. Figures 2 and 3 are examples of GINGER and GENESIS simulations, respectively. The main difference between GINGER and GENESIS is that GINGER assumes an axisymmetric radiation beam, while the radiation field in GENESIS is fully three dimensional. Simulation algorithms based on the integral solution of the paraxial wave equation have been used for RON [46] and FAST [47] to reduce the computation time and to aid design optimization.

IV SCHEMES TO ENHANCE PERFORMANCE

Self-amplified spontaneous emission is the most straightforward approach to achieve extremely high-brightness x-ray sources. Various schemes have been proposed to improve the performance of the x-ray FELs by employing more sophisticated magnetic and optical configurations. We review some of them here.

A High-Gain Harmonic Generation

Although SASE FELs have excellent transverse modes, poor temporal coherence is a limitation imposed by the shot noise buildup because the coherence length $c\sigma_\tau = c/(2\sigma_\omega)$ is usually much less than the bunch length. Without a proper coherent seed at x-ray wavelengths, a high-gain harmonic-generation (HGHG) FEL resorts to a coherent seed at longer wavelengths. In this scheme [48], a small energy modulation is imposed on the electron beam by interaction with a seed laser in a short undulator (the modulator). The energy modulation is converted to a coherent spatial density modulation as the electron beam traverses a dispersive section. A second undulator (the radiator), tuned to a higher harmonic of the seed frequency, causes the microbunched electron beam to emit coherent radiation at that harmonic frequency, followed by exponential amplification until saturation is achieved. The HGHG output radiation has a single phase determined by the seed laser, and its spectral bandwidth is Fourier transform limited [49].

A crucial condition in HGHG FELs is that the beam energy spread entering the radiator section must be made much less than the FEL parameter ρ, in order for the harmonic bunching at the dispersive section to be effective. A successful demonstration experiment has been carried out at Brookhaven National Laboratory (BNL) [49,50], in which a seed CO_2 laser at a wavelength of 10.6 μm produced a saturated, amplified FEL output at the second-harmonic wavelength, 5.3 μm. The intensity of nonlinear harmonic radiation at 2.65 μm and 1.77 μm were determined relative to that of the 5.3-μm fundamental [50]. Cascading several stages of HGHG FELs have been envisioned in order to reach shorter wavelengths [51].

B Regenerative Amplifier

A regenerative amplifier FEL or RAFEL consists of a high-gain undulator and a narrow-band optical feedback system [52]. A small fraction of the SASE signal amplified from shot noise is monochromatized and back-reflected as the input signal for the subsequent amplifications. If mirrors of reasonable reflectivity exist at a certain wavelength range, the steady-state, temporally coherent signal can be achieved after only a few passes of operation. For example, such a scheme has been proposed at the DESY FEL facility at VUV wavelengths ranging from 60 nm to 200 nm [53].

C   Two-Stage SASE FEL

Another novel scheme to improve the temporal coherence is the so called two-stage SASE FEL [54], which consists of two undulators (of the same undulator period and strength) and an x-ray monochromator located between them. The first undulator operates in the linear regime of a SASE FEL. After the exit of the first undulator, the electron is guided through a bypass, and the SASE signal enters the monochromator, which selects a narrow band of radiation. At the entrance of the second undulator the monochromatic x-ray beam is combined with the electron beam and is amplified up to the saturation level. Since the SASE power over a narrow bandwidth at the exit of the first undulator fluctuates 100 percent according to the exponential probability distribution, the length of the second undulator is chosen to exceed the saturation length sufficiently to suppress fluctuation of the final output power level.

D   X-ray Pulse Compression

The length of an x-ray SASE pulse cT is normally determined by the electron beam, which is about 100 fs. However, it consists of x-ray wavelets, each about $(\rho)^{-1}$ periods, where the FEL scaling parameter $\rho \sim 10^{-3}$. Therefore, it should be possible to compress the SASE pulse to the coherent length at saturation $(c\tau)_{\text{coh}} \approx \lambda/\rho$, which is about 1 fs for 1 Å [55]. The compression is accomplished by introducing an energy slew in the electron beam leading to the frequency chirp (frequency shift per unit length). The maximum chirp that can be imposed without degrading the FEL gain is $\frac{\rho}{\lambda/\rho} = \rho^2/\lambda$. The total correlated energy spread in the beam is $(\Delta\omega/\omega)_{\text{max}} = \rho^2 T/\lambda$. The chirped pulse can then be compressed with a grating pair. The minimum pulse length after compression given by phase-space conservation is $(c\tau)_{\text{min}} = \lambda/\rho$, which is the same as the coherence length at saturation. The technique for pulse compression has been extensively developed for high-power solid state lasers at visible wavelengths. For the x-ray SASE pulse, a major research project will be to demonstrate that the required optical elements exist.

E   Variable Polarization

A crossed undulator configuration was proposed for a high-gain FEL for versatile polarization control [56]. It consists of a long (saturation length) planar undulator, a dispersive section, and a short (a couple of gain lengths) planar undulator oriented perpendicular to the first one. In the first undulator, a radiation component linearly polarized in the x-direction is amplified to saturation. In the second undulator, the x-polarized component propagates freely, while a new component polarized in the y-direction is generated from the bunched beam and reaches the same intensity by coherent spontaneous emission. By adjusting the strength of the dispersive section, the relative phase of the two radiation components can be adjusted to obtain a

TABLE 2. Typical parameters of recent SASE FEL experiments.

Main parameters	UCLA/LANL	LEUTL	TTF	VISA
Beam energy [MeV]	18	217	233	71
Energy spread	0.25%	0.1%	0.15%	0.18%
Peak current [A]	170	150	400	200
Normalized emittance [μm]	4	7.5	6	2
Betatron wavelength [m]	1.2	9	4.5	1.8
Undulator period [cm]	2.05	3.3	2.73	1.8
Undulator parameter	1.04	3.1	1.2	1.2
Effective undulator length [m]	2	21.6	13.5	4
FEL wavelength	12 μm	530 nm	109 nm	0.8 μm

suitable polarization, including the circular polarization, for the total radiation field.

V   SURVEY OF PROOF-OF-PRINCIPLE EXPERIMENTS

Until recently, FEL experiments operated in the SASE mode have been in the millimeter wavelength range [57]. Driven by intense R&D efforts towards x-ray FELs, a large number of recent SASE FEL experiments push the lasing wavelengths from infrared to visible and on to ultraviolet. We review some of the most significant SASE experiments with the typical parameters given in Table 2.

In the UCLA/LANL experiment [58], radiation intensity at 12 μm is increased by more than 10^4 when the electron charge is changed by a factor of seven (0.3-2.2 nC). Using the measured beam parameters at 2.2 nC, a power gain of 3×10^5 is deduced. The intensity fluctuations are well described by the gamma function distribution with the M parameter evaluated from experimental data.

In the ongoing APS LEUTL experiment [59], exponential gains at 530 nm and 385 nm are measured along the nine undulator sections. With the measured beam parameters, the fitted gain length, near and far-field mode sizes, and radiation spectrum are found to agree with SASE theories. The LEUTL experiment has extensive electron, radiation beam diagnostics (as shown in Fig. 6), capable of measuring SASE evolution as a function of undulator distance z. The longitudinal microbunching of the electron beam along z is observed using coherent transition radiation [60]. With the operation of a bunch compressor, intensity amplification of more than 10^5 is measured, and saturation at both wavelengths is achieved (see Fig. 7). More data collection and analysis are planned at these wavelengths and at shorter wavelengths such as 120 nm.

In the DESY TTF experiment [61], electron beams accelerated by a superconducting rf linac are used to obtain the shortest FEL wavelengths to date. Initial

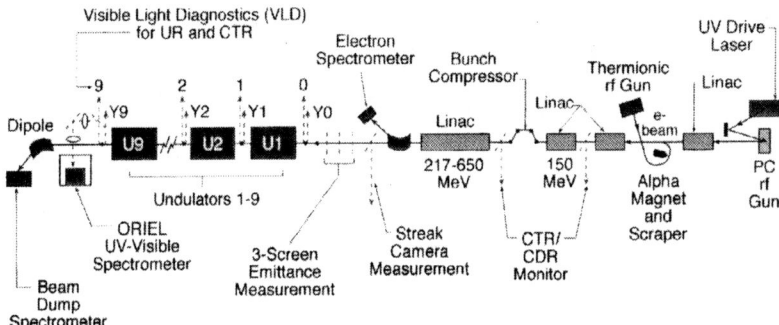

FIGURE 6. Schematic of the electron and radiation beam diagnostics at LEUTL (courtesy of A. Lumpkin, ANL).

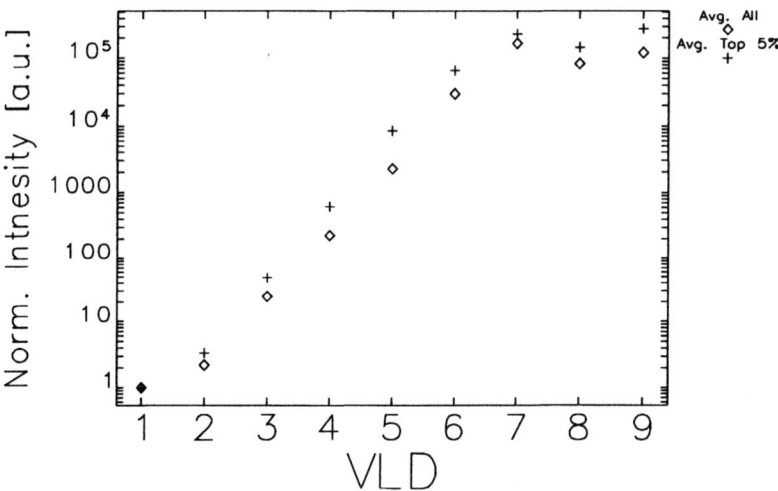

FIGURE 7. Measurements of exponential growth and saturation of 530 nm SASE FEL at LEUTL (each VLD station corresponds to a 2.4-m length of undulator, courtesy of S. Milton, ANL).

results show a gain of about 10^3 from 180 nm to 80 nm, and the radiation characteristics, such as dependency on bunch charge, angular distribution, spectral width, and intensity fluctuation, are all consistent with the present models for SASE FELs.

The VISA group at BNL is commissioning a 0.8-0.6 μm experiment, using a 4-m-long undulator with distributed strong focusing quadrupoles. Initial results show a gain of about 100 [62].

VI   CONCLUSIONS

We presented the theoretical advances during the last decade and recent successes in demonstration experiments of high-gain free electron lasers in SASE and other modes of operation for generation of extremely high brightness x-rays. With these developments, the x-ray sciences will have the opportunity to experience another advance as revolutionary as the development of the synchrotron radiation source.

REFERENCES

1. Y.S. Debenev, A.M. Kondratenko, and E.L. Saldin, Nucl. Instrum. Methods Phys. Res. A **193**, 415 (1982).
2. R. Bonifacio, C. Pellegrini and L.M. Narducci, Opt. Commun. **50**, 373 (1984).
3. Linac Coherent Light Source Design Study Report, SLAC-R-521, 1998.
4. LCLS, the first experiments, September 2000.
5. D.E. Moncton, Proceedings of XIX International LINAC Conference, Chicago, IL, USA, 1048 (1998).
6. Conceptual Design of a 500 GeV e+e- Linear Collider with Integrated X-ray Laser Facility, DESY 97-048, 1997.
7. J.S. Fraser, R.L. Sheffield, and E.R. Gray, Nucl. Instrum. Methods Phys. Res. A, **250**, 71 (1986).
8. B.E. Carlsten, Nucl. Instrum. Methods Phys. Res. A, **285**, 313 (1989).
9. D.T. Palmer *et al.*, SPIE **2522**, 514 (1995).
10. P. Emma and R. Brinkmann, Proceedings of the 1997 Particle Accelerator Conference, Vancouver, BC, Canada, 1679 (1997). V.K. Bharadwaj *et al.*, *ibid*, 903 (1997).
11. Ya.S. Derbenev, J. Rossback, E.L. Saldin, and V.D. Shiltsev, DESY Report No. TESLA-FEL-95-05, 1995.
12. T. Raubenheimer, Nucl. Instrum. Methods Phys. Res. A **358**, 40 (1995).
13. E. Gluskin *et al.*, to be published in the Proceedings of the 22nd International Free Electron Laser Conference, Durham, NC, USA, 2000.
14. K.-J. Kim, Nucl. Instrum. Methods Phys. Res. A **250**, 396, (1986).
15. J.-M. Wang and L.-H. Yu, *ibid*, 484 (1986).
16. K.-J. Kim, Phys. Rev. Lett. **57**, 1871 (1986).
17. S. Krinsky and L.-H. Yu, Phys. Rev. A **35**, 3406 (1987).
18. L.-H. Yu, S. Krinsky, and R.L. Gluckstern, Phys. Rev. Lett. **64**, 3011 (1990).
19. Y.H. Chin, K.-J. Kim, and M. Xie, Phys. Rev. A **46**, 6662 (1992).

20. M. Xie, Nucl. Instrum. Methods Phys. Res. A **445**, 59 (2000).
21. M. Xie, presented at the American Physical Society April Meeting, Long Beach, 2000 and to be published in the Proceedings of the 22nd International Free Electron Laser Conference, Durham, NC, USA, 2000.
22. Z. Huang and K.-J. Kim, to be published in the Proceedings of the 22nd International Free Electron Laser Conference, Durham, NC, USA, 2000.
23. M. Xie, Proceedings of the 1995 Particle Accelerator Conference, 183 (1995).
24. E.L. Saldin, E.A. Schneidmiller, and M.V. Yurkov, Nucl. Instrum. Methods Phys. Res. A **313** 555 (1992); see also *The Physics of Free Electron Lasers*, Springer, Berlin, 1999.
25. L.-H. Yu and S. Krinsky, Nucl. Instrum. Methods Phys. Res. A **285**, 119 (1989).
26. N.A. Vinokurov, Z. Huang, O.A. Shevchenko, and K.-J. Kim, to be published in the Proceedings of the 22nd International Free Electron Laser Conference, Durham, NC, USA, 2000.
27. R.W. Warren, J.C. Goldstein, and B.E. Newnam, Nucl. Instrum. Methods Phys. Res. A **250**, 19 (1986).
28. M. Xie, Nucl. Instrum. Methods Phys. Res. A **445**, 67 (2000).
29. E.L. Saldin, E.A. Schneidmiller and M.V. Yurkov, submitted to Opt. Commun., 2000.
30. Z. Huang and K.-J. Kim, to be published.
31. R. Bonifacio *et al.*, Phys. Rev. Lett. **73**, 70 (1994).
32. J.W. Goodman, *Statistical Optics*, John Wiley & Sons, New York, 1985.
33. E.L. Saldin, E.A. Schneidmiller, and M.V. Yurkov, Opt. Commun. **148**, 383, 1998.
34. K.-J. Kim, in *Towards X-ray Free Electron Lasers*, edited by R. Bonifacio and W.A. Barletta, AIP, New York, 1997.
35. R. Bonifacio, L. De Salvo, and P. Pierini, Nucl. Instrum. Methods Phys. Res. A **293**, 627 (1990).
36. H.P. Freund, S.G. Biedron, and S.V. Milton, IEEE J. Quantum Electron. **QE-36**, 275 (2000); Nucl. Instrum. Methods Phys. Res. A **445**, 53 (2000).
37. Z. Huang and K.-J. Kim, Phys. Rev. E **62**, 7259 (2000).
38. K.-J. Kim, in *Quantum Aspects of Beam Physics*, edited by P. Chen, World Scientific, Singapore, 1999.
39. C. Schroeder, C. Pellegrini, and P. Chen, to be published in the Proceedings of the 2nd Workshop on Quantum Aspects of Beam Physics, Capri, Italy, 2000.
40. H.-D. Nuhn, SPIE **3614**, 119 (1999).
41. S. Biedron *et al.*, Nucl. Instrum. Methods Phys. Res. A **445**, 110 (2000).
42. T.-M. Tran and J.S. Wurtele, Computer Physics Comm. **54**, 263 (1989)
43. W.M. Fawley, CBP Tech Note-104, Lawrence Berkeley Laboratory, 1995.
44. S. Reiche, Nucl. Instrum. Methods Phys. Res. A **429**, 243 (1999).
45. C. Penman and B.W.J. McNeil, Opt. Commun. **90** 82 (1992).
46. R.J. Dejus, O.A. Shevchenko, N.V. Vinokurov, Nucl. Instrum. Methods Phys. Res. A **429**, 225 (1999).
47. E.L. Saldin, E.A. Schneidmiller, and M.V. Yurkov, Nucl. Instrum. Methods Phys. Res. A **429**, 225 (1999).
48. L.H. Yu, Phys. Rev. A **44**, 5178 (1991).

49. L.-H. Yu *et al.*, Science, **289**, 932 (2000).

50. A. Doyuran *et al.*, to be published in the Proceedings of the 22nd International Free Electron Laser Conference, Durham, NC, USA, 2000.

51. J. Wu and L.-H. Yu, submitted for publication, 2000.

52. J.C. Goldstein, D.C. Nguyen, and R.L. Sheffield, Nucl. Instrum. Methods Phys. Res. A **393**, 137 (1997).

53. B. Faatz *et al.*, Nucl. Instrum. Methods Phys. Res. A **429**, 424 (1999).

54. J. Feldhaus *et al.*, Opt. Commun. **140**, 341 (1997).

55. C. Pellegrini, Nucl. Instrum. Methods Phys. Res. A **445**, 124 (2000).

56. K.-J. Kim, Nucl. Instrum. Methods Phys. Res. A **445**, 329 (2000).

57. T. Orzechowski *et al.*, Phys. Rev. Lett. 54, 889 (1985); D. Kirkpatrick *et al.*, Nucl. Instrum. Methods Phys. Res. A **285**, 43 (1989).

58. M.J. Hogan *et al.*, Phys. Rev. Lett. **81**, 4867 (1998).

59. S.V. Milton *et al.*, Phys. Rev. Lett. **85**, 988 (2000); N.D. Arnold *et al.*, to be published in the Proceedings of the 22nd International Free Electron Laser Conference, Durham, NC, USA, 2000.

60. A.H. Lumpkin *et al.*, Phys. Rev. Lett. **86**, 79 (2001).

61. J. Andruszkow *et al.*, Phys. Rev. Lett. **85**, 3825 (2000).

62. A. Tremaine *et al.*, to be published in the Proceedings of the 22nd International Free Electron Laser Conference, Durham, NC, USA, 2000.

HIGH-GAIN FREE-ELECTRON LASERS AND HARMONIC GENERATION

Sandra G. Biedron[†+], Giuseppe Dattoli[*], Henry P. Freund[‡], Zhirong Huang[†], Stephen V. Milton[†], Heinz-Dieter Nuhn[#], Pier Luigi Ottaviani[°], Alberto Renieri[*]

[†] *Advanced Photon Source, Argonne National Laboratory, Argonne, IL 6043, USA*

[*] *ENEA, Divisione Fisica Applicata, Centro Ricerche Frascati, C.P. 65, 00044 Frascati, Rome, Italy*

[‡] *Science Applications International Corporation, McLean, VA 22102, USA*

[#] *Stanford Linear Accelerator Center, Stanford University, Menlo Park, CA 94309-0210, USA*

[°] *ENEA, Divisione Fisica Applicata, Centro Ricerche Bologna, Bologna, Italy*

[+] *and MAX Laboratory, University of Lund, Lund, Sweden S-22102*

Abstract. We consider a free-electron laser (FEL) operating in the high-gain regime including the mechanism of self-induced harmonic generation (SIHG). We discuss the mechanism leading to saturation at these higher harmonics and analyze the effect of beam quality variations. Finally, we study an undulator configuration that allows the saturated harmonic power to become comparable with that of the fundamental.

INTRODUCTION

The mechanisms behind coherent, self-induced harmonic generation (SIHG) in free-electron laser (FEL) devices has been the subject of both older and more recent theoretical [1-7] and experimental activity [8-9]. SIHG is a promising method to obtain shorter wavelengths in, for an example, self-amplified spontaneous emission (SASE) FELs, since not only the fundamental but the higher harmonics achieve saturation nearly simultaneously. This occurs in all single-pass, high-gain, free-electron lasers configured with planar undulators. The sinusoidal electron beam traversal through these undulators naturally forces the odd harmonics to be favored.

The possibility of exploiting SIHG to coherently generate higher-order harmonics, which eventually may be exploited to 1) themselves serve as a radiation source or 2) serve as a seed for FEL operation at shorter wavelengths, has been a leitmotif for the design of devices operating in the VUV-X region of the spectrum [4-7,9-14].

CP581, *Physics of, and Science with, the X-Ray Free-Electron Laser,* edited by S. Chattopadhyay et al.
© 2001 American Institute of Physics 0-7354-0022-9/01/$18.00

In this contribution we discuss several schemes of high-gain FELs exploiting SIHG and, in particular, we will deal with the effect of the beam quality on higher harmonics and possible methods to enhance the output power of the higher harmonics.

Let it be known that we are merely stating what has been discussed in the September 2000 Arcidosso Workshop, "The Physics of, and Science with, the X-ray Free-Electron Laser" in Arcidosso, Italy; in a follow-up meeting held in November at the Advanced Photon Source (APS) at Argonne National Laboratory (ANL); and in many subsequent other forms of contact. We do not make any attempt to provide an unanimous conclusion as some of us still have a difference in opinion.

DISCUSSION OF THE NUMERICAL RESULTS

We will consider an FEL operating in the SASE regime with the parameters listed in Table 1. The undulator and electron beam parameters have been chosen in accordance with those of Ref. [3] and we have checked that the numerical method employed in the present investigation, based on a one-dimensional macroparticle code, PROMETEO, is capable of reproducing results similar to the three-dimensional simulation analysis of Ref. [3] after an appropriate choice of the electron beam current density (see Table 1).

In Fig. 1 (a-d) we plot the evolution of the first three odd harmonics for different values of the electron beam relative energy spread. In the case of Fig. 1 (a) ($\sigma_\varepsilon = 5 \times 10^{-4}$), all harmonics reach saturation and the saturated power of the fundamental is larger than the SIHG by one to two orders of magnitude.

It is well known that less desirable electron beam quality affects the fundamental and, therefore, harmonic evolution. The linear harmonics have been previously shown to be significantly affected by a degraded electron beam quality [15]. The nonlinear harmonics appear to be less sensitive to energy spread with respect to the fundamental as shown in Fig. 1 (b, c, d), where we have considered $\sigma_\varepsilon = 10^{-3}, 1.5 \times 10^{-3}, 2 \times 10^{-3}$, respectively. The numerical results clearly display that, while the evolution of the first harmonic (fundamental) is not significantly distorted by an increase of the energy spread, the SIHG exhibits a mild reduction of the saturated power. Using PROMETEO, the saturation of the fundamental is always followed by that of SIHG, at least for the range of parameters we have explored.

TABLE 1. List of parameters

E (MeV) = 219.5
I (A) = 150
e-beam transverse area (m^2) = 3.8×10^{-7}
Current density (A/m^2) = 3.95×10^{8} A/m^2
ε_n (mm-mrad) = 5π
λ_u (cm) = 3.3
K = 3.1

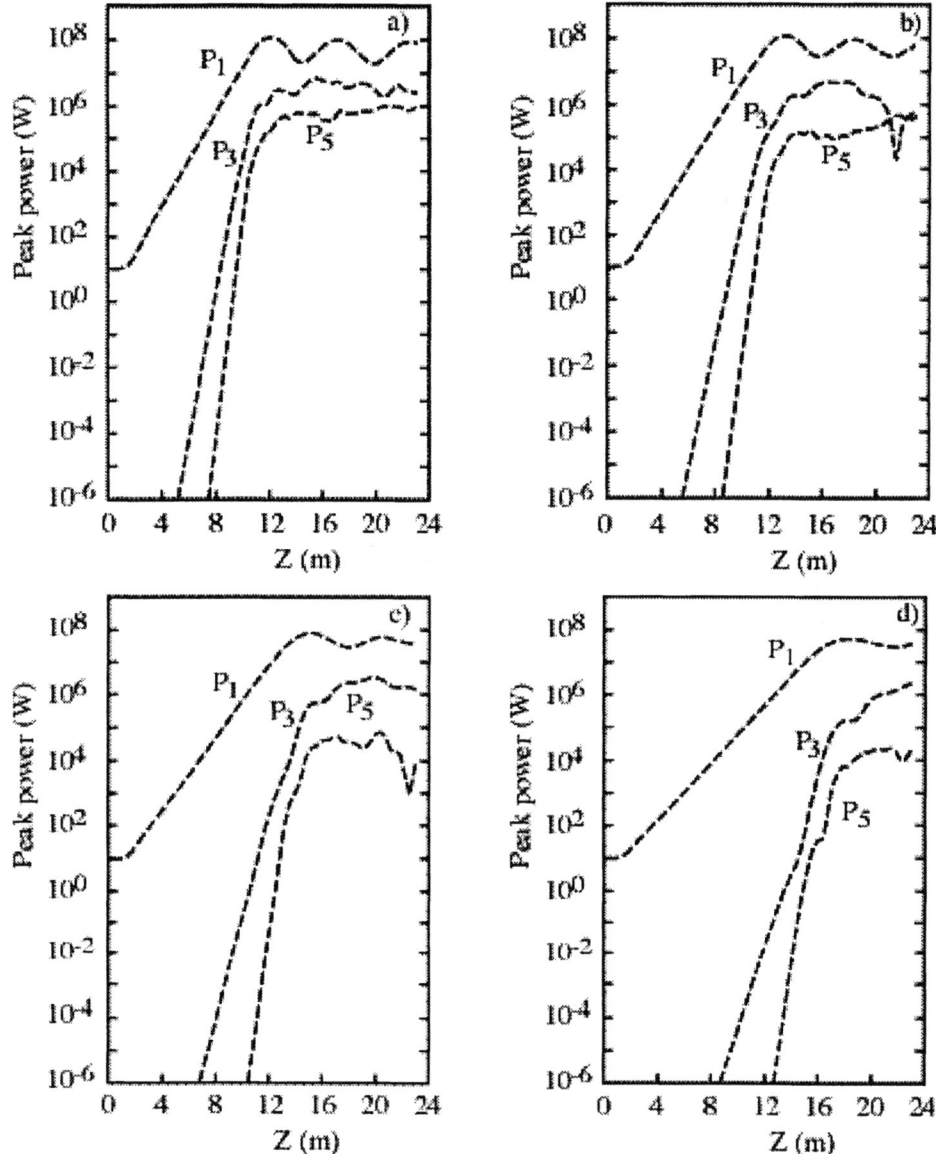

FIGURE 1. Evolution of first three odd harmonics vs z(m). a) $\sigma_\varepsilon = 5 \times 10^{-4}$, b) $\sigma_\varepsilon = 10^{-3}$, c) $\sigma_\varepsilon = 1.5 \times 10^{-3}$, d) $\sigma_\varepsilon = 2 \times 10^{-3}$.

To further analyze these results, we have performed parameter scans of the electron beam energy spread and emittance for the same parameter set as shown in Table 1 using the three-dimensional macroparticle computer code known as MEDUSA, as discussed in reference [3]. These scans were made using the fundamental and third harmonic only. In this analysis, the powers of the fundamental and third harmonic are chosen at the point of the fundamental saturation. In Figs. 2 and 3, the variation of both the fundamental and third harmonic output power is shown as a function of the energy spread and emittance, respectively. Note that as the energy spread or emittance degrades, the harmonic power decreases slightly more rapidly than that of the fundamental. This is not so much, however, that the harmonic power would become unusable.

The choice made for the current density in the one-dimensional code PROMETEO differs from that found in MEDUSA. MEDUSA follows the propagation of both the electromagnetic field and a matched beam self-consistently in three dimensions for a given emittance and, for the case under study, yields an rms beam radius of about 180 μm corresponding to a current density of about $1.5 \times 10^9 \text{A/m}^2$. Hence, the results may differ, and the comparison between the two codes may change if the higher current density would be applied in a PROMETEO simulation run.

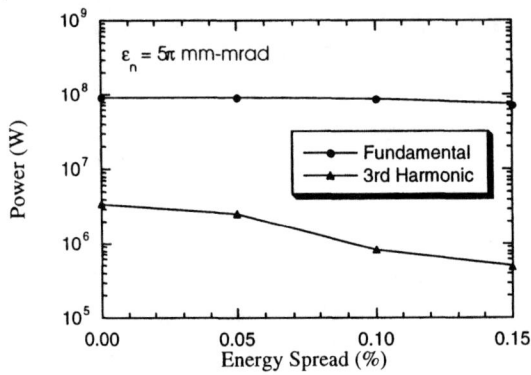

FIGURE 2. Power (W) of fundamental and third harmonic as a function of the electron beam energy spread (%).

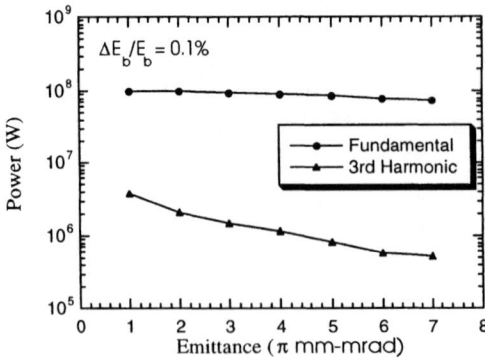

FIGURE 3. Power (W) of fundamental and third harmonic as a function of the electron beam emittance (π mm-mrad).

A more effective method of comparing this deviation is to examine the ratio of the harmonic power to the fundamental. For the PROMETEO analysis, this is shown in Fig. 4. For the MEDUSA analysis, the data for the harmonic-to-fundamental ratios as a function of energy spread and emittance are shown in Figs. 5-6.

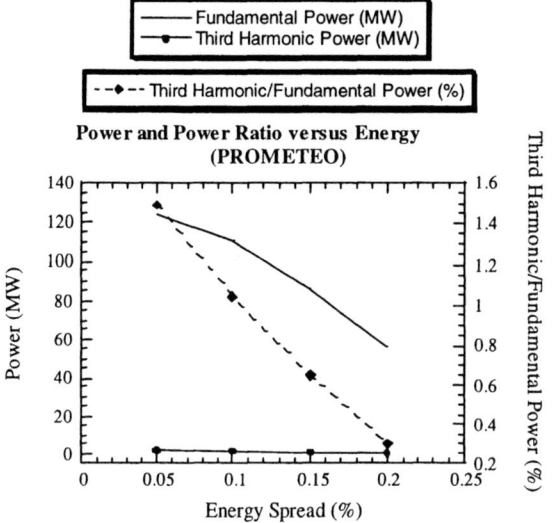

FIGURE 4. Fundamental and third harmonic power (W) and power ratio (%) of the third harmonic to the fundamental as a function of the electron beam energy spread (%) for the 1-dimensional macroparticle case.

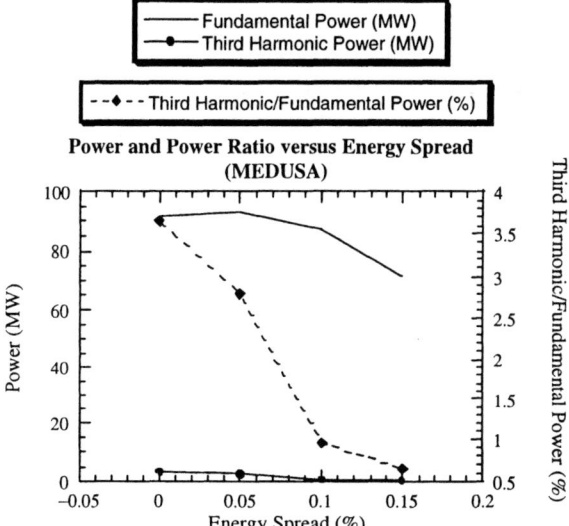

FIGURE 5. Fundamental and third harmonic power (W) and power ratio (%) of the third harmonic to the fundamental as a function of the electron beam energy spread (%) for the MEDUSA analysis.

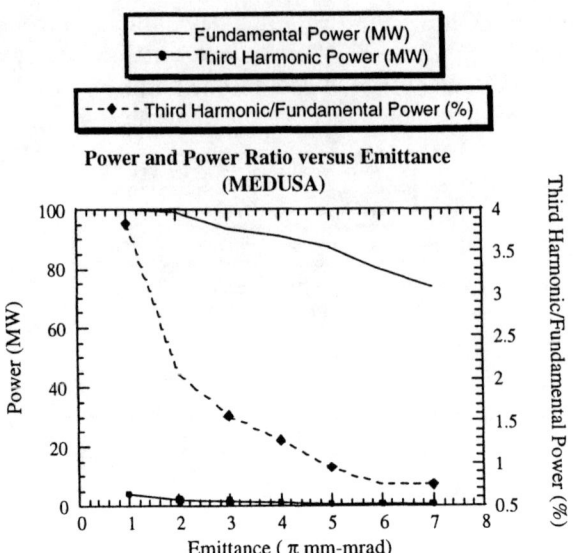

FIGURE 6. Fundamental and third harmonic power (W) and power ratio (%) of the third harmonic to the fundamental as a function of the electron beam emittance (π mm-mrad) for the MEDUSA analysis.

In the previous simulations, only the nonlinear harmonics produced by SIHG using a single undulator were followed. In such SIHG, the power of the fundamental is always significantly larger than that of the harmonics. Generally speaking, the third harmonic power at saturation is in the range of one percent of the fundamental.

We now will analyze an undulator configuration to enhance the power of the SIHG. The undulator consists of two parts: the first with the parameters given in Table 1, the second with a period exactly 1/3 of that of the first section but with the same K parameter. In this example the third harmonic of the first undulator corresponds to the fundamental of the second undulator and the electron beam prebunching induced in the first part may provide the condition for the growth of the power at $\lambda/3$ in the second section.

Figure 7 shows the results for different values of the energy spread. We have considered different lengths of the first section and have found that the *higher harmonics* may reach power levels comparable with those of the fundamental for small values of the energy spread. When σ_ε exceeds 10^{-3}, the gain length becomes prohibitively long in the second section. The case not presented here places the second undulator just after the third harmonic reaches saturation. Also note that the amount of power that can be generated by tuning the second undulator to the third harmonic is due to the coefficient of the bunching parameter, the coupling parameter, that goes as $(K_1/K_h)^2$, where $K_h = K(-1)^{(h-1)/2}\left[J_{(h-1)/2}(h\xi) - J_{(h+1)/2}(h\xi)\right]$ and $\xi = K^2/(4 + 2K^2)$. So, since the desired SIHG wavelength is now the fundamental of the second undulator, the amount of power available increases because the ratio becomes one.

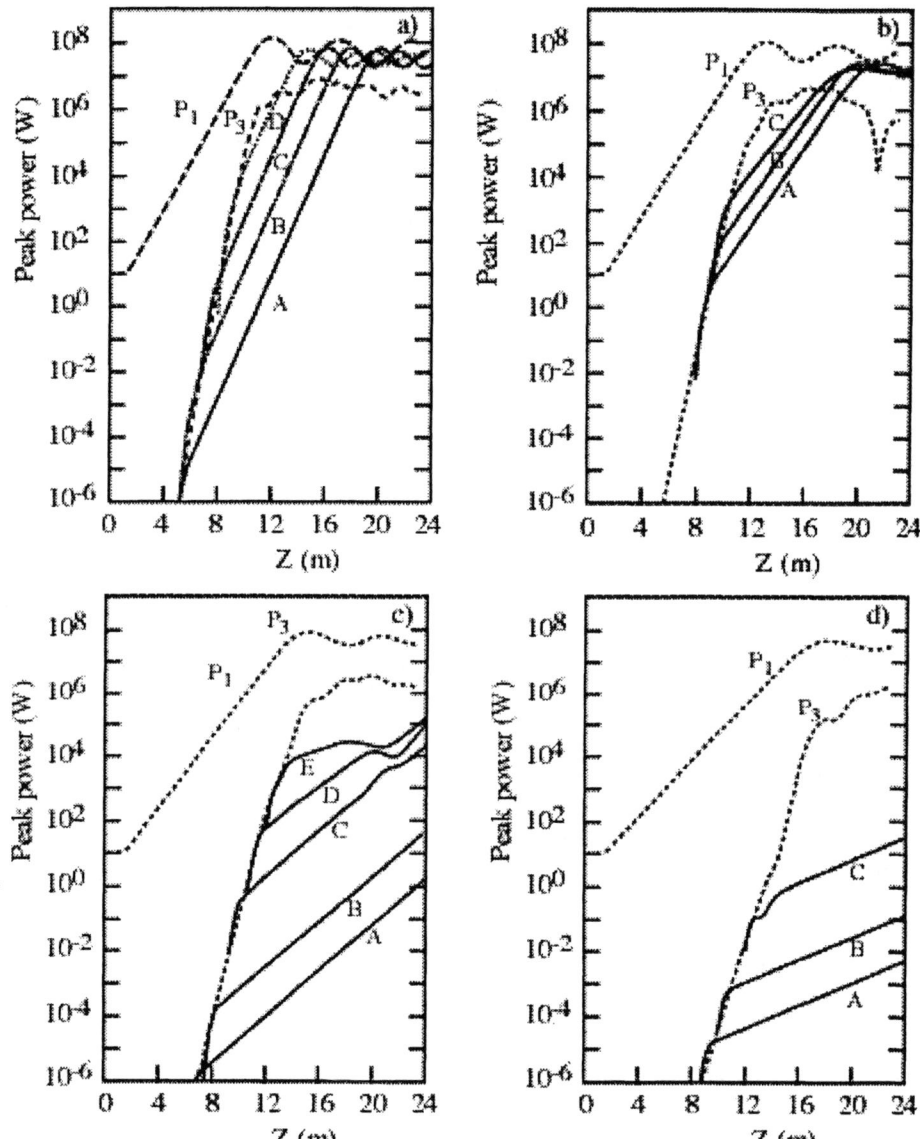

FIGURE 7. a) Evolution of the first and third harmonics as in Fig. 1 (a) ($\sigma_\varepsilon = 5 \times 10^{-4}$). The curves A, B, C, D refer to the case of composite undulator configuration, the section with $\lambda_u/3$ starts after a number of periods A) N = 120, B) N = 160, C) N = 200, D) N = 240. (b) Same as (a) ($\sigma_\varepsilon = 10^{-3}$), A) N = 240, B) N = 260, C) N = 280. (c) Same as (a) ($\sigma_\varepsilon = 1.5 \times 10^{-3}$), A) N = 180, B) N = 220, C) N = 280, D) N = 320, E) N = 360. (d) Same as (a) ($\sigma_\varepsilon = 2 \times 10^{-3}$), A) N = 260, B) N = 300, C) N = 360.

FINAL REMARKS

In this note we have explored some aspects of SIHG in high-gain FELs as were discussed in the Arcidosso workshop, in the follow-up meeting at the APS, and in subsequent communications. We have drawn the following conclusions:

♦ degrading electron beam quality produces a mildly worse effect on the SIHG power than on the fundamental but does not eliminate their overall usefulness,

♦ undulators with different period lengths can be exploited to enhance the power in a desired SIHG,

♦ a proper electron beam quality is crucial for the successful operation of an FEL scheme.

The reäson why the two-undulator system cannot reach saturation in a reasonable length of the second undulator is under analysis and will be discussed more carefully elsewhere.

Finally, this group of individuals have agreed to continue meeting informally every few months to further discuss and critique exotic source schemes and issues regarding user facilities based upon FEL theory and experiment.

ACKNOWLEDGEMENTS

The work of S. G. Biedron, Z. Huang, and S. V. Milton were supported at Argonne National Laboratory by the U.S. Department of Energy, Office of Basic Energy Sciences under Contract No. W-31-109-ENG-38. The activity and computational work for H. P. Freund was supported by Science Applications International Corporation's Advanced Technology Group under IR&D subproject 01-0060-73-0890-000. The work of H.-D. Nuhn was supported by the U.S. Department of Energy, Office of Basic Energy Sciences, Division of Material Sciences, under contract No. DE-AC03-76SF00515.

REFERENCES

1. Colson, W.B., Dattoli, G., and Ciocci, F, *Phys. Rev.* **31A**, 828-842 (1985).
2. Benson, S.V. and Madey, J.M.J., *Phys. Rev.* **39A**, 1579-1581 (1989).
3. Schmitt, M.J., *Phys. Rev.* **41A**, 3853-3866 (1990).
4. Bonifacio, R., de Salvo Souza, L., Pierini, P., and Scharlemann, E.T., *Nucl. Instrum. Methods in Phys. Res.* **A296**, 787 (1990).
5. Fawley, W.M., Nuhn, H.-D., Bonifacio, R., Scharlemann, E.T., *Proc. IEEE 1995 Particle Accelerator Conf.*, 219-221 (1995).
6. Freund, H.P., Biedron, S., and Milton, S.V., *IEEE J. Quant. Electron.* **36**, 275-281 (2000).
7. Huang, Z. and Kim, K.-J., *Phys. Rev.* **62E**, 7295-7308 (2000).
8. Prazeres, R. et al. *Nucl. Instrum. Methods in Phys. Res.* **429A**, 131-135 (1999).
9. Doyuran, A., Babzien, M., Shaftan, T., Biedron, S.G., Yu, L.-H. et al. *Nucl. Instrum. Methods in Phys. Res. A*, accepted and in press, to appear in 2001.
10. Ben Zvi, I. et al., *Nucl. Instrum. Methods in Phys. Res.* **318A**, 208-211 (1992).
11. Ciocci, F. et al., *IEEE J. Quant. Electron.* **31**, 1242-1252 (1995).
12. Dattoli, G. and Ottaviani, P. L., *J. Appl. Phys.* **86** 5331-5336 (1999).
13. Biedron, S.G., Milton, S.V., and Freund, H.P., *Nucl. Instrum. Methods in Phys. Res. A*, accepted and in press, to appear in 2001.
14. Yu, L.-H. and Wu, J., *Proceedings of the 22nd Free Electron Laser Conference and 7th FEL Users Workshop*, 13-20 August 2000, Durham, North Carolina, submitted.
15. Bonifacio, R., Corsini, R., and Pierini, P., *Phys. Rev.* **45A**, 4091-4096 (1992).

An optical experiment towards the analogic simulation of the betatronic motion

A.Bazzani[†], P.Freguglia[⋆],L.Fronzoni[∗], G.Turchetti[†]

† Dept. of Physics Univ. of Bologna and INFN Sezione di Bologna, ⋆ Dept. of Pure and Applied Mathematics Univ. of L'Aquila and Domus Galilaeana (Pisa), ∗ Dept. of Physics Univ. of Pisa, INFM and CISSC of the Univ. of Pisa

ABSTRACT

We present the first experiment results of an experiment that simulates the betatronic motion in a FODO cell by using an optical device. The analogy between the hamiltonian mechanics and the geometric optics is discussed in order to define a project of an optical FODO cell. We consider also the possibility of representing an optical path by means of a sequence of quaternion numbers. The experimental results are compared with the simulations of a ray tracing program.

INTRODUCTION

The analogy between the geometric optics and the mechanical system has suggested the hamiltonian formulation of Mechanics and the possibility to derive a wave equation for a particle[1]. Moreover it is interesting to remember, for instance, that Jean Bernoulli (see Solution problematis a se in Actis 1696 p.269 propositi) and A.F. Mbius (see Lehrbuch der Statik, 1837), as Ernst Mach says us in its Die Mechanik in ihrer Entwickelung historisch-kritisch dargestellt (1883), notice useful analogies "respectively between motion of masses and equilibrium of threads and between motion of masses and motion of the light". This analogy has suggested an approach to the betatronic dynamics by a wave equation, whose solutions open new experimental prospects for the accelerator physics. We want to use the same analogy to consider the feasibility of an optical experiment, which enables the simulation of some aspects of nonlinear dynamics. In particular we are interested in the simulation of the betatronic motion in circular particle accelerators[2]. The nonlinear effects due to the magnetic multipoles in the lattice of an accelerator can introduce long term instabilities in the transverse beam dy-

CP581, *Physics of, and Science with, the X-Ray Free-Electron Laser,* edited by S. Chattopadhyay et al.
© 2001 American Institute of Physics 0-7354-0022-9/01/$18.00

namics that decrease the luminosity. Some experiments have been performed in order to measure the dynamic aperture and the nonlinear effects in the betatronic motion of a real accelerator. These experiments are extremely complex and expensive. The tracking programs for the particle dynamics usually contain some approximation and their results do not always explain the experimental data. An optical device that simulates the betatronic dynamics in a FODO cell is not expensive and could be used as analogic simulator. Moreover it could check the effect of the various approximation introduced in the tracking program and to analyze the transport. The geometric optics approximation of the light propagation is presented in the section 2. by deriving the eikonal equation from the wave equation of the electromagnetic field that implies the Fermat's variational principle for the ray trajectories and the hamiltonian formulation of the equation of motion. Then we compute the dynamics of a light ray through the discontinuity surfaces of a thick lens by using the Snell's law.

In section 3. we consider the equivalence between an optical system with two spherical lenses of opposite convexity and a FODO cell in a particle accelerator and we show that a spherical lens introduces in a first approximation a nonlinear octupolar like effect in the ray dynamics.

In section 4. we consider a purely geometrical description of the trajectory by utilizing a sequence of quaternions whose application to the geometric optics has been suggested by Hamilton[3].

In section 5. we discuss the problems connected with the project of an optical FODO cell and we compare the preliminary experimental data with the results of a ray tracing program.

GEOMETRIC OPTICS APPROACH TO THE LIGHT PROPAGATION

The wave equation derived from the Maxwell theory of the electromagnetic field reads

$$\nabla^2 \mathbf{v} - \frac{n^2}{c^2} \ddot{\mathbf{v}} = 0 \tag{1}$$

where $\mathbf{v}$ can be identified both with the electric and magnetic field and we have introduced the refractive index n of the medium. In the geometric optics approximation one looks for a solution of eq. (1) in the form

$$\mathbf{v}(\mathbf{x}, t) = \mathbf{v}(\mathbf{x}) e^{i k_0 S(\mathbf{x})} e^{i \omega t} \tag{2}$$

where $\mathbf{v}(\mathbf{x})$ is a real vector related to the wave amplitude, the real function $S(\mathbf{x})$ defines the phase of the wave and $k_0 = \omega/c = 2\pi/\lambda_0$, λ_0 being the vacuum wavelength. Then one considers the limit $\lambda_0 \to 0$ and reduces the wave equation (1) to

$$\mathbf{v} \left(\|\nabla S\|^2 - n^2 \right) = 0 \tag{3}$$

A nontrivial solution exists only if S satisfies the eikonal equation

$$\|\nabla S\|^2 - n^2 = 0 \tag{4}$$

and the surfaces of constant phase $S(\mathbf{x}) = const.$ define the wavefronts.
The geometric optics approximation holds if the wavelength λ_0 is much smaller than the spatial derivatives of the wave amplitude. The geometrical light rays are normal to the wave fronts and satisfy the equation

$$\frac{d\mathbf{x}}{ds} = \frac{\nabla S}{n} \tag{5}$$

where s is the arc length parameter of the trajectory. A simple calculation shows that $ds = dS/n$: i.e. the distance between two wavefronts is inversely proportional to the refractive index. Then one defines the optical path between two points P_1 and P_2 as

$$[P_1 P_2] = \int_{P_1}^{P_2} n \, ds = S(P_1) - S(P_2) \tag{6}$$

where the integral is computed along a light ray; the optical path is proportional to the traveling time. If $n(\mathbf{x})$ has no singularity in the space, the equation (5) implies that the integral

$$\oint_{\gamma(P_1 P_2)} n \frac{d\mathbf{x}}{ds} \cdot d\mathbf{x} \tag{7}$$

does not depend on the particular path γ connecting the points P_1 and P_2; the equation (7) is equivalent to the Fermat variational principle that characterizes the light rays as the extremals of the functional

$$\int_{\gamma(P_1 P_2)} n \, ds \tag{8}$$

The Fermat principle proves the symplectic character of the dynamics in the geometric optics approximation. Indeed if we introduce the generating function

$$F(\mathbf{x}_1, \mathbf{x}_2) = \int_{\gamma(P_1 P_2)} n \, ds \tag{9}$$

where $\mathbf{x}_i$ are the coordinates of the point P_i and $\gamma(P_1 P_2)$ is the light ray connecting the two points, then the definition (6) and the equation (5) give

$$n \frac{d\mathbf{x}_2}{ds} = \frac{\partial F}{\partial \mathbf{x}_2} \qquad n \frac{d\mathbf{x}_1}{ds} = -\frac{\partial F}{\partial \mathbf{x}_1} \tag{10}$$

Therefore by introducing the momentum $\mathbf{p} = n \, d\mathbf{x}/ds$, the phase flux defined by the ray dynamics in the phase space $(\mathbf{x}, \mathbf{p})$ is symplectic.

The variational principle (9) allows to describe the crossing of a discontinuity surface Σ of the refractive index $n(\mathbf{x})$ by a light ray. If $\mathbf{p}_1$ and $\mathbf{p}_2$ are the momenta associated to the ray trajectory just before and after the discontinuity surface Σ, the extremality condition of the functional (9) implies the relation

$$\mathbf{p}_1 \cdot d\mathbf{x} = \mathbf{p}_2 \cdot d\mathbf{x} \tag{11}$$

for are virtual displacement $d\mathbf{x}$ tangent to the surface Σ at the incident point. Therefore the tangent component of the momentum $\mathbf{p}$ is continuous across the surface Σ, whereas the normal component is determined by the constraint $\|\mathbf{p}\| = n$. This is the content of the Snell's law. If one explicitly knows the normal versor $\hat{\nu}$ to the surface Σ at the incident point, the equation (11) gives the relation

$$\mathbf{p}_2 = \mathbf{p}_1 - (\mathbf{p}_1 \cdot \hat{\nu})\hat{\nu} + \sqrt{(\mathbf{p}_1 \cdot \hat{\nu})^2 + n_2^2 - n_1^2}\, \hat{\nu} \tag{12}$$

where n_1 (resp. n_2) is the value of the refractive index just before (resp. after) the surface Σ. If we consider an optical device in which the refractive index is piecewise constant then the trajectory of a light ray is rectilinear in the homogeneous regions and changes direction according to eq. (12) at the discontinuity surfaces. However the relation (12) does not define a symplectic map in the canonical coordinates $(\mathbf{x}, \mathbf{p})$ since the optical path of the light ray is different for different incident points.

DYNAMICS OF A FODO SYSTEM OF OPTICAL LENSES

An optical device analog to the FODO cell structure of the particle accelerators is built by two parallel spherical lenses of opposite convexity separated by empty spaces. We introduce a coordinate system $\mathbf{x}$ by choosing the straight line passing through the lens centers as z-axis (reference orbit). We consider the sections defined by the planes tangent to the surfaces of each lens and perpendicular to the z-axis: the position of the k-section is determined by its z_k. Moreover we recall that $\|\mathbf{p}\| = n$ and the p_z component can be computed from the other components. As a consequence the dynamical state of a light ray is defined by the phase space coordinates (x, y, p_x, p_y) in each section. The dynamics of a light ray through the device can be described by introducing the symplectic maps $\mathcal{M}_k$ that define the correspondence between the phase space coordinates associated to two nearby sections. We distinguish two kind of maps: the maps associated to an empty section and the maps associated to the passage through a lens. In the first case we have the relations

$$\begin{aligned} \mathbf{x}_{k+1} &= \mathbf{x}_k + s\mathbf{p}_k \\ \mathbf{p}_{k+1} &= \mathbf{p}_k \end{aligned} \tag{13}$$

where s is a parameter determined by the z-component of eq. (13) as a function of the distance $z_{k+1} - z_k$ between two nearby sections. In the second case the map is defined according to eq. (12) by computing the incident points of the ray at the surfaces of the lens (see fig. 1). We explicitly perform the calculation of the transfer map for a spherical lens of curvature radius R in the limits $x, y \ll R$. This approximation is analog to the approximation used to derive the usual equation for betatronic motion in a particle accelerator. In order to preserve the symplectic character of the transfer map, we assume that the rays are almost parallel to the z-axis and we neglect the contribution of p_x, p_y in the r.h.s. of eq. (12). In this case the transverse coordinates (x, y) of the incident ray do not change crossing the lens (thin lens approximation) and the transfer map can be computed by developing the r.h.s. of eq. (12) for $x, y \ll R$. Since the p_z component is determined by the constrain $\|p\| = n$ we restrict our computation to the transverse components and the transfer map of a focusing spherical thin lens reads

$$
\begin{aligned}
p'_x &= p_x - 2(n-1)\frac{x}{R} - \frac{n^2-1}{n}\frac{x(x^2+y^2)}{2R^3} \\
p'_y &= p_y - 2(n-1)\frac{y}{R} - \frac{n^2-1}{n}\frac{y(x^2+y^2)}{2R^3}
\end{aligned}
\tag{14}
$$

where n is the refractive index of the material (we assume that the refractive index in the empty space is equal to 1). The equation (14) takes into account the first nonlinear contribution that comes from the expansion of eq. (12) and shows that the spherical optical lens introduces an octupolar-like effect in the ray dynamics.

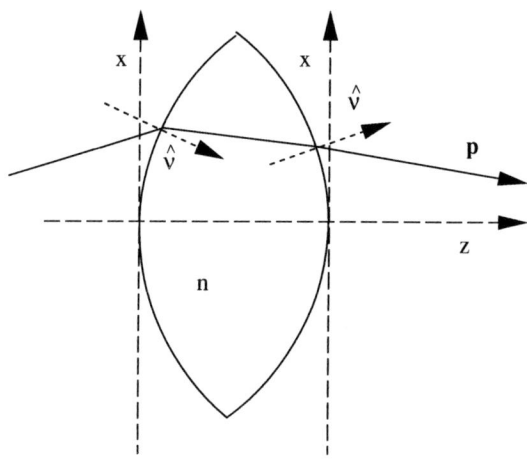

Figure 1: optical path of a light ray through a thick lens.

In a circular accelerator the FODO cells are introduced to confine the particle in a neighborhood of the reference closed orbit. Therefore to simulate the dynamics of a particle we have to consider several passages of the light ray through the optical FODO: this can be achieved by using a system of mirrors (see next section) that close the reference trajectory through the lens centers. In this case the dynamics is studied by introducing the Poincarè map of the system that is the result of the composition of the transfer maps of the elements and gives the intersections of the ray orbits in the phase space with a fixed transverse section (Poincarè section). The linear part of the Poincarè map $\mathcal{M}$ describes the dynamics of the orbits near the reference orbit and we choose the parameters of the optical FODO so that in the linear approximation $\mathcal{M}$ can be reduces to a rotation around an elliptic fixed point. In the Poincarè section the orbits move on invariant ellipses and rotate by a fixed angle (tune) each iteration. However according to eq. (14), the dynamics of the Poincarè map turns out to be a nonlinear and the phase space shows the effects of nonlinear resonances and chaotic orbits.

APPLICATION OF THE QUATERNION NUMBERS TO THE DESCRIPTION OF THE ORBITS.

In the previous sections we have discussed the dynamical approach to the study of the evolution of a ray through an optical device; however since we would like to use the optical FODO as a analogic simulator we are interested also in developing a purely geometrical description of the orbit that could used to study the geometrical properties from the experimental data[4]. In the thin lens approximation we can associate a transverse section to the position of each lens along the reference orbit. Then the optical path of a light ray is given by a sequence of vectors that connect the intersecting points of each section (see fig. 2).

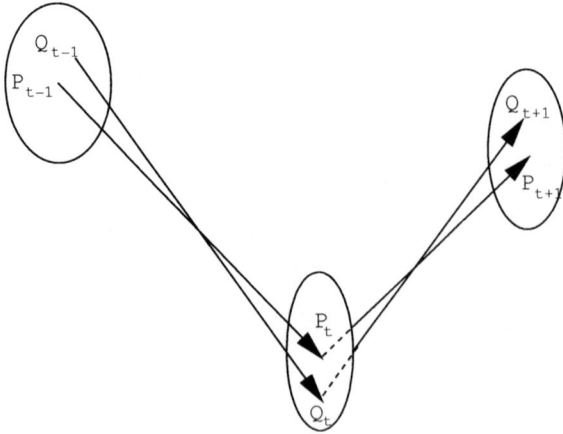

Figure 2: geometrical description of the light propagation.

The vectors $P_t P_{t+1} = \mathbf{u}_t$ define the reference orbit and are considered known. The sequence of vectors $Q_t Q_{t+1} = \mathbf{w}_t$ defines the optical path of a generic light ray in the device. Following Hamilton[3] we associate a quaternion number to each vector according to

$$\mathbf{u} \to \mathbf{u} \cdot \mathbf{i} \qquad (15)$$

where $\mathbf{i} = (i, j, k)$ are the imaginary units of the quaternion space; then it is possible to define the sequence of quaternion

$$\mathbf{r_t} = \mathbf{w}_t \mathbf{u}_t^{-1} = \frac{(-\mathbf{u}_t \cdot \mathbf{w}_t + (\mathbf{u}_t \times \mathbf{w}_t) \cdot \mathbf{i})}{\|\mathbf{u}_t\|^2} \qquad (16)$$

as the "right ratio" of the vectors $\mathbf{w}_t$ and $\mathbf{u}_t$. The sequence $\mathbf{r}_t$ gives a quaternionic representation of the orbit of a generic light ray and takes into account the effect of the lenses according to the Snell's law.

Another representation of an optical path can be derived by means of a "Λ"-determinability procedure[5,6]. If we fix a point O in the space, an orbit can also be determined by the vectors $OQ_t = \mathbf{v}_t$ that satisfy the relation $\mathbf{v}_{t+1} = \mathbf{v}_t + \mathbf{w}_t$. Then we introduce a sequence of quaternions $\mathbf{q}_t$ according to

$$\mathbf{q}_t \mathbf{v}_t = \mathbf{v}_{t+1} \qquad (17)$$

which is related to the intersection of the orbit at each section; we have a quaternionic relation between $\mathbf{r}_t$ and $\mathbf{q}_t$

$$(\mathbf{q}_t - 1)\mathbf{v}_t = \mathbf{r}_t \mathbf{u}_t \Rightarrow \mathbf{q}_t = \mathbf{r}_t \mathbf{u}_t \mathbf{v}_t^{-1} + 1 \qquad (18)$$

The study of the geometrical properties of the sequences $\mathbf{q}_t$ and $\mathbf{r}_t$ in the quaternionic space will be considered in the next future.

THE EXPERIMENTAL SETUP

We have simulated the quadrupole magnets by using two cylindrical lenses L_1 and L_2 of opposite convexity with a radius of curvature .6 m and diameter .05 m. The refractive index of the glass is 1.6 and corresponds to a focal distance of .5 m. The distance between the two lenses is .66 m so that the linear dynamics is characterized by a tune of $\omega = .23$. In order to create the closed reference orbit we use 2 mirrors (see fig. 3).

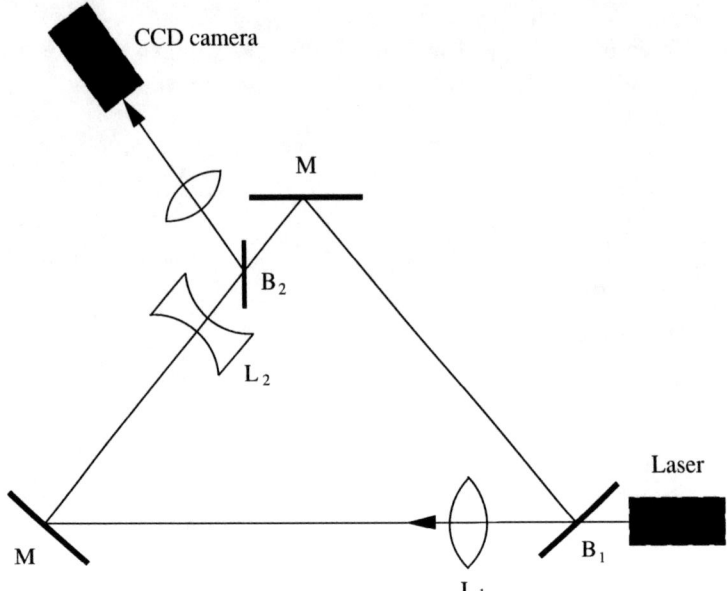
Figure 3: scheme of the experimental setup of an optical FODO.

The injected light is obtained from a 50 mW He-Ne laser by means of a beam splitter B_1 whose transmission coefficient should be very small to minimize the beam losses (.1 is the actual value). The laser gives a thin, high intensity and well collimated beam which is required to compensate the beam losses in the device, but it introduce interference effect since the laser light is coherent.

The position of the light beam is detected by using a beam splitter B_2 with a small reflection index (.34 is the actual value) which deflects part of the beam on the screen of a CCD camera through a focusing lens L_3. Because of the presence of the optical loop many spots appear in the camera screen. The order of the succession of the points is inversely proportional to the intensity of the spots. The reconstruction of the phase space of the orbit is achieved by measuring the beam position at two different sections at a distance of .02 m.

The apparatus results a good simulator of the betatronic motion if many turns of the beam are detectable, but we have to avoid the spurious effects due to:

a) the reflection of the beam at the lenses surfaces;

b) the duplication of the beam due to the lenses;

c) the loss of energy due to the absorption of the lenses;

d) the irregularity of the mirror surfaces.

The point a) and b) could be avoided with a special treatment of the lens surface, but more easily it is enough to rotate the lenses of a small angle respect to the axis of the reference orbit. This deflection allows to put away the reflections of the lenses. The points c) and d) require a choice of a suitable material with a suitable treatments of the glasses surfaces.

In the present state only a small numbers of spots are detectable and we can perform a qualitative analysis of the experimental data. In fig. 4 we compare the preliminary experimental data with a numerical simulations which consider thick lenses in the ray tracking. We remark that the motion is almost linear in the considered region of the phase space. In the figure 5 we plot a larger region of the phase space of the Poincarè map in the flat beam approximation obtained from the numerical simulations; the presence of the nonlinear effects are clearly visible at a distance from the reference orbit of 3 cm in the camera screen. Therefore we expect to see some nonlinear effect in the next future after an upgrade of the lenses quality.

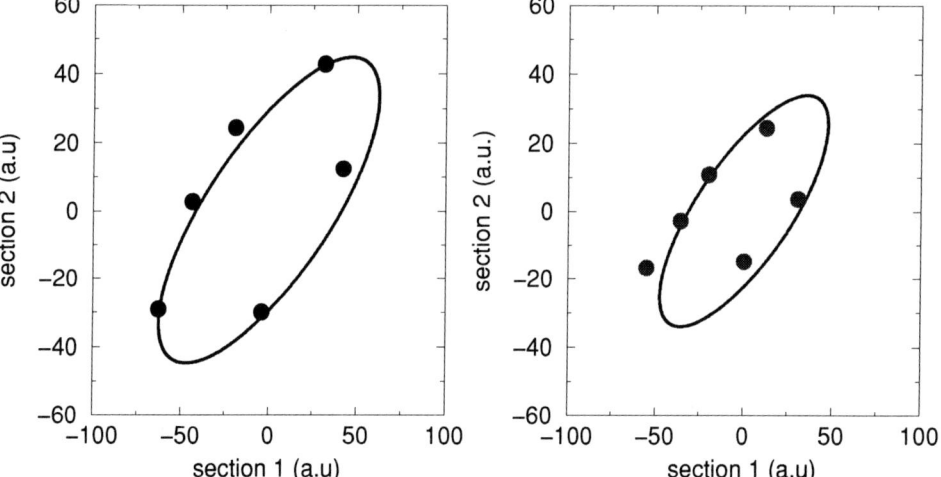

Figure 4: comparison of the experimental data (filled circles) with the invariant curves obtained from a numerical ray tracing program (continuous lines) for two different initial conditions (the unity is related to the arbitrary unity read from the screen of the CCD camera).

CONCLUSIONS

The analogy between hamiltonian systems and geometric optics suggested the idea to construct an optical device which results an "analogic simulator" to study the nonlinear effects in the betatronic motion in the circular accelerators. This first experiment has the purpose to study the feasibility of such a device and points out the advantages and the problems of an optical simulator of a nonlinear system. At the same time we have considered the possibility of representing an optical path by using the a sequence of quaternion numbers that could help in the analysis of the geometrical properties of the orbit.

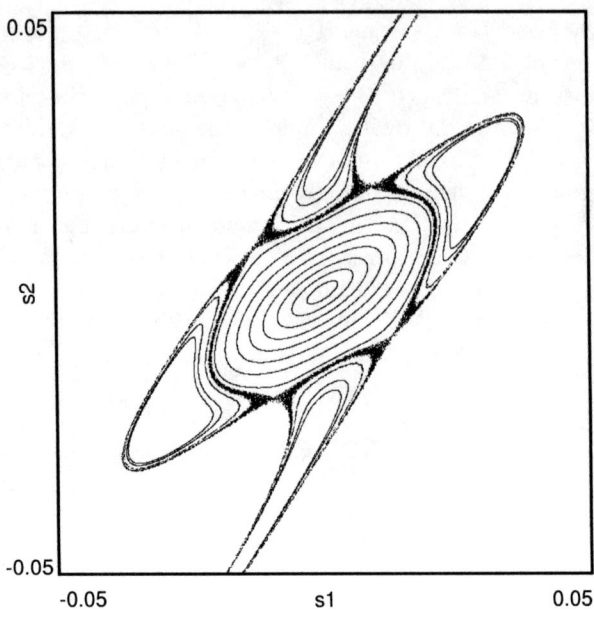

Figure 5: phase space of the Poincarè map computed by the ray tracing program that simulates the experiment; the horizontal (resp. vertical) axis reports the beam horizontal positions at the first (resp. second) section $s1$ (resp. $s2$) considered in the experiment: the scales are in meters.

REFERENCES

1. M.Born,E.Wolf, *Principles of Optic*,Cap. III e IV,Pergamon Press, London, (1959).

2. A.Bazzani,E.Todesco,G.Turchetti, "*A normal form approach of nonlinear betatronic motion*",**CERN 94-02**, (1994).

3. W.R.Hamilton *Lectures on quaternions*, Dublin, (1853).

4. A.Bazzani, P.Freguglia, "*An optical geometric model of the betatronic motion*", in *The application of mathematics to the sciences of nature: critical moments and aspects*, Kluwer Academic-Plenum Publishers, New York, (2001).

5. P. Freguglia ,"*A synthetic geometrical approach to betatronic motion: some remarks*", *IOP Conference Proceedings* ,vol. **468**, (1999).

6. Ronald E.Mickens, "*Difference Equations Theory and Applications*",ed. Van Nostrand Reinhold, New York, (1990).

Optimization of an X-ray SASE-FEL

C. Pellegrini*, S. Reiche*, J. Rosenzweig*, C. Schroeder*,
A. Varfolomeev†, S. Tolmachev† and H.-D. Nuhn**

*UCLA – Dept. of Physics & Astronomy
†Russian Research Center Kurchatov Institue
**Stanford Linear Accelerator Laboratory

Abstract. The most important characteristics of an X-ray SASE-FEL are determined by the electron beam energy, transverse and longitudinal emittance, and by choice of the undulator period, field, and gap. Among them are the gain and saturation length, the amount and spectral characteristics of the spontaneous radiation, the wake fields due to the vacuum pipe. The spontaneous radiation intensity is very large in all X-ray SASE-FELs now being designed, and it contributes to the final electron beam energy spread, thus affecting the gain. It also produces a large background for the beam and radiation diagnostics instrumentation. The wake fields due to the resistivity and roughness of the beam pipe through the undulator, also affects the beam 6-dimensional phase space volume, and thus the gain and the line width. In this paper we discuss ways to optimize the FEL when considering all these effects. In particular we consider and discuss the use of a hybrid iron-permanent magnet helical undulator to minimize some of these effects, and thus optimize the FEL design.

INTRODUCTION

When moving towards shorter FEL wavelengths the demands on the electron beam parameters and undulator design becomes more severe. While the typical undulator length are longer, the tolerable emittance and energy spread are smaller. Therefore the FEL performance is more sensitive to effects such as wake fields and spontaneous emission.

The emission of incoherent undulator radiation is the same for all electrons, and yields an energy loss, which can be compensated by tapering the undulator, and an increased energy spreads, which can reduce the gain. Wake fields effects change depending on the electron position within the bunch, and produce an energy loss from the entrance to the exit of the undulator which reduces the gain.

The wake fields effects can however be reduced by proper design of the vacuum chamber, increasing its diameter and reducing the resistivity and the roughness, and of the undulator, to reduce its length by maximizing the gain. The general strategy to reduce wake fields effects is to increase the vacuum chamber diameter and to shorten the undulator length. Therefore in this discussion we compare the planar LCLS undulator, considered as a reference, with helical undulators, in particular a novel design by the Kurchatov Institute for a helical undulator with very high magnetic field. With this design a shorter length can be achieved compared to planar undulator.

CP581, *Physics of, and Science with, the X-Ray Free-Electron Laser,* edited by S. Chattopadhyay et al.
© 2001 American Institute of Physics 0-7354-0022-9/01/$18.00

TABLE 1. Parameters for LCLS (planar undulator) and alternative models based on helical undulators (cases A – C)

	LCLS	A	B	C
Undulator period, cm	3	3	4	3
Undulator field, T	1.3	0.96	0.48	0.96
Undulator K	3.7	2.7	1.8	2.7
Undulator gap, mm	6.0	8.5	8.0	8.5
Focusing beta function, m	18.0	17.7	20.5	5.0
Beam energy, GeV	14.4	14.7	10.65	14.7
Total synchrotron radiation, GW	90	50	11.6	10
Normalized emittance, mm mrad	1.1	1.1	1.1	0.3
Charge, nC	0.95	0.95	0.95	0.2
Peak current, kA	3.4	3.4	3.4	1.17
Relative energy spread at undulator entrance, 10^{-5}	6	6	8	6
Resonant wavelength, Å	1.5	1.5	1.5	1.5
FEL parameter, $\rho \times 10^4$	5	6	6	10
Gain Length, m	4.2	2.8	4.2	1.84

UNDULATOR AND BEAM PARAMETERS MODELS

Both planar and helical undulators, have been successfully used for FELs in the past. In the models, presented in this paper, we consider helical undulators with large gaps, except for the LCLS reference case, which uses a planar hybrid undulator. The choice of high field helical undulator is motivated by the fact that with a new design helical undulators can give a short gain length, while the gap can be as large as in the planar case. With a larger gap the vacuum chamber size can be increased, thus reducing the impact of wake fields (see Sec.).

For the discussion we consider 4 cases:

- a high field helical undulator (case A);
- a low field helical undulator (case B);
- a low charge, high field helical undulator (case C);
- the reference LCLS case, as presented in the CDR [1].

The exceptionally high field helical undulator is a permanent magnet system designed by the Kurchatov group. This undulator can provide a large field or a large gap (see Fig. 1). The choice has to be determined by considerations of wake field effects in the undulator vacuum pipe, and total undulator length. For this discussion a gap width of 8.5 mm is used providing the resonant wavelength at almost the same energy as the LCLS undulator.

In case B a different, conventional design for a helical undulator is used, providing a lower magnetic field and larger period length. This requires a lower electron beam energy to obtain the same resonance wavelength. The benefits of the model is the rather simple design of the undulator. Beside combinations of permanent magnets and iron yokes the same field profile can be obtained by a double helix of current carrying copper embedded in an iron yoke. The required current does not exceed 2000 A, and the Ohmic

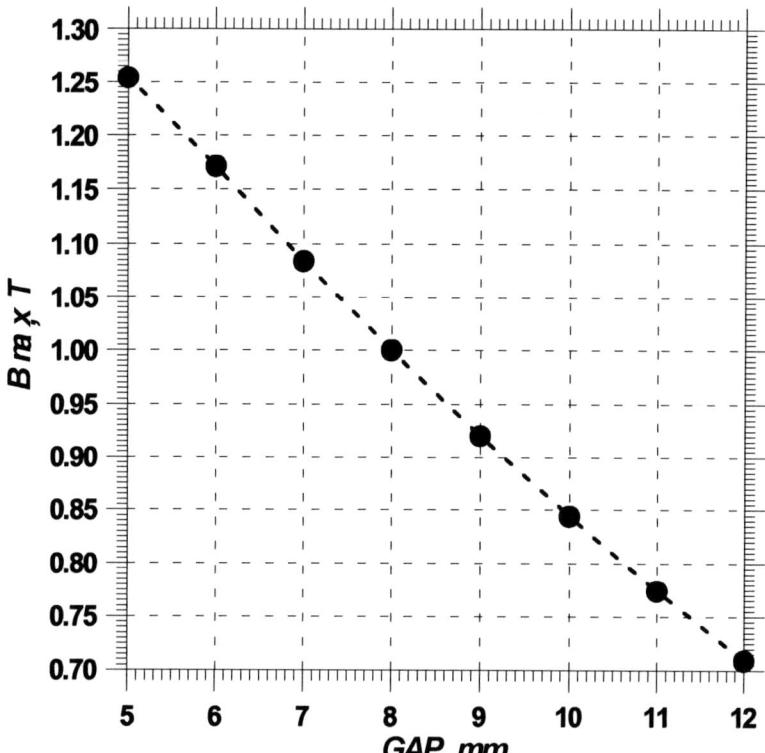

FIGURE 1. On-axis Field strength versus gap for the Kurchatov design of the helical undulator.

losses can be cooled by liquid nitrogen.

Model C is a modified version of Model A, where the bunch charge and, thus, the bunch length and beam emittance are smaller [2]. The reduced emittance effects and the use of stronger focusing results in a shorter undulator.

The undulator and electron beam parameters for the four cases are presented in Tab. 1.

For completeness a fifth model could have been considered, using a planar undulator with larger gap while providing the same on-axis peak field as the LCLS case. The construction of this undulator was suggested by Kurchatov group also. The physics would have been same as for the LCLS case except that the wake fields are reduced to the level of the helical models. This impact is covered by Sec. .

POWER AND SATURATION LENGTH

For the simulation we used the 3D time-dependent FEL code GENESIS 1.3 [3], which has been benchmarked to various other FEL codes in the steady-state regime of an FEL

[4] . To reduce the CPU time and to exclude many independent runs to exclude the fluctuation due to the SASE process all models have been simulated as an FEL amplifier. The initial power level has been estimated by the 1D theory [5], which is applicable for X-ray FELs because they are not dominated by diffraction effects.

The general performance of the different models is given by the solid lines in Fig. 2 (Sec. discusses the wake fields effect, also shown in this plot).

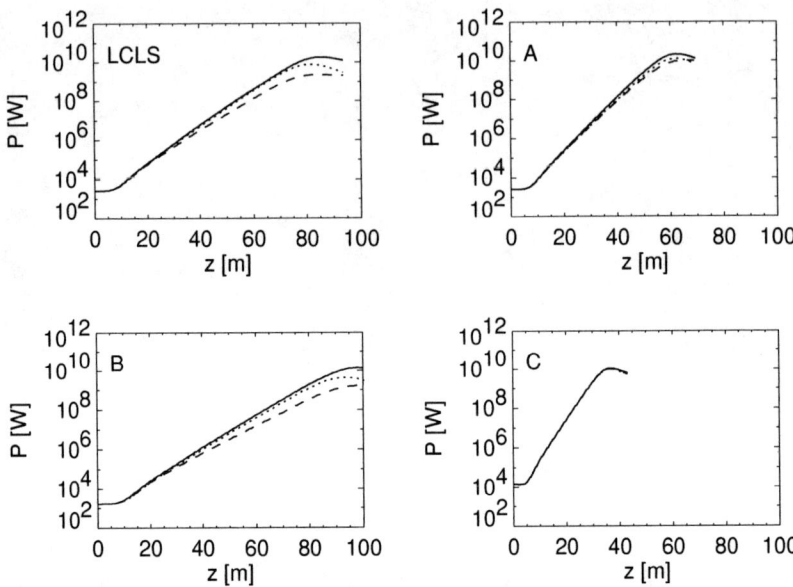

FIGURE 2.   Radiation power for the different cases including no wake fields, resistive wall wake fields and resistive wall + surface roughness wake fields (solid, dotted and dashed line, respectively)

For all models the saturation level is almost identical at 10 GW while the saturation length varies. Model C shows the best performance with a saturation length of 37 m due to the reduced emittance and stronger focusing. But even with the same electron beam parameters the performance of the high field undulator (case A) exceeds that of the LCLS undulator. The low field helical undulator has the longest saturation length. This is caused by the emittance effects which are stronger for a lower beam energy while keeping the undulator parameter the same [6].

In the conceptional design study of LCLS the ability to tune the resonant wavelength between 1.5 Åand 15 Åis an important feature for a wider range of experiments with a high brilliant x-ray beams. All models are able to fulfill this requirement by changing only the electron beam energy. At 15 Åthe saturation length is below 30 m for all cases. The benefits of a reduced bunch charge and emittance are not as significant as for the 1.5 Åcase, because the FEL performance is less affected by emittance effects at lower beam energy.

SPONTANEOUS EMISSION

The difference between a planar and a helical undulator becomes obvious when the spontaneous radiation is taken into account. Typically a planar undulator radiates at a higher power lever with a richer content of higher harmonics than a helical undulator. Because the FEL is driven by a lower beam energy case B has the lowest radiation level considered in this discussion. In addition spontaneous radiation of a helical undulator is always emitted at an angle with respect to the undulator axis, which makes it easier to separate the FEL radiation from the spontaneous emission.

Another effect for X-ray FELs is the fluctuation in the number of emitted photons in the high frequency part of spectrum of the radiation spectrum. It yields an increased energy spread of the electron beam [7], which cannot be compensated by field tapering as it is the case for the average energy loss due to the spontaneous emission.

The growth rate of the energy spread scales with the electron beam energy. Therefore case B benefits from the lower beam energy and the FEL is hardly affected. For the other cases the initial energy spread grows about 100% before it is dominated by the saturation of the FEL (see Fig. 3). Because the FEL dynamic is rather affected by the electron beam emittance than the energy spread the overall effect of the quantum fluctuation in the spontaneous emission is small. Even for the worst case the power level at saturation is degraded by less than 10%.

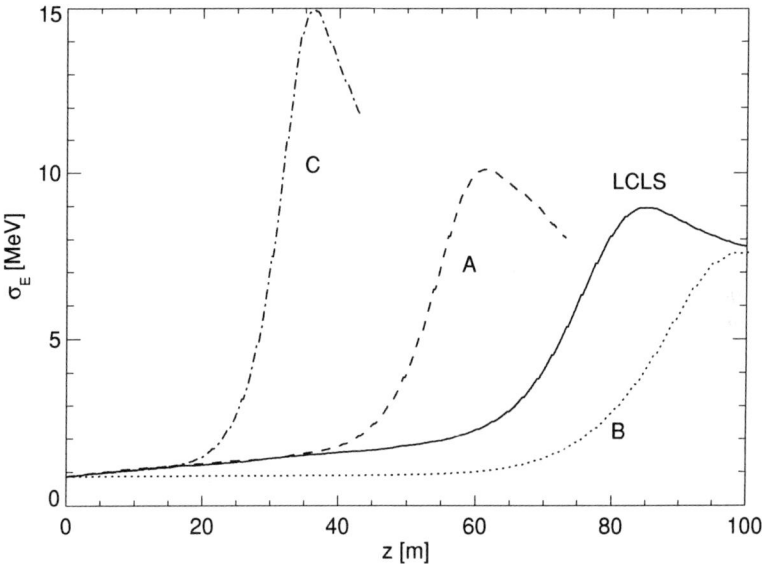

FIGURE 3. Energy spread along the undulator including the effect due to the fluctuation of the emitted photons in the spectrum of the spontaneous emission.

WAKE FIELDS EFFECTS

Wake field effects within the undulator cannot be neglected due to the high peak current of the electron beam and the small diameter of the vacuum chamber, which enhance the wake field amplitude significantly. These effects have been the subject of much recent work [8, 9, 10]. The two main effects considered are those due to the resistivity and the imperfections of the vacuum pipe wall. A third source of wake field, produced by discontinuities in the vacuum chamber, is excluded in this discussion because an estimate of the wake field amplitude relies on an explicit design of the vacuum chamber including pumping ports, diagnostic sections and bellows.

Wake fields have longitudinal and transverse components. The latter are not taken into account because they are of higher order in the electron beam misplacement. Longitudinal wake fields cause a modulation of the electron beam energy along the bunch, and are commonly described by a wake potential.

For the resistive wall wake fields the wake potential [11] is

$$W_z(t) = -\frac{4ce^2Z_0}{\pi R^2}\left[\frac{1}{3}e^{t/\tau}\cos(\sqrt{3}t/\tau) - \frac{\sqrt{2}}{\pi}\int_0^\infty dx\frac{x^2e^{tx^2/\tau}}{x^6+8}\right] \qquad (1)$$

where t is the longitudinal position of the test particle with respect to the particle generating the field, $\tau = [2R^2/Z_0\sigma c^3]^{1/3}$ is the characteristic scale of the wake potential, σ is the conductivity of the vacuum chamber, R is the chamber radius and $Z_0 \approx 377\ \Omega$ is the vacuum impedance.

For the effect of imperfection there are several models are under consideration. A more conservative model [8] describes the surface roughness by a thin dielectric layer where the thickness δ is equivalent to the rms surface modulation.

The resulting wake potential is

$$W_z(t) = -\frac{ce^2Z_0}{\pi R^2}\cos(k_0 t) \qquad (2)$$

with $k_0 \approx \sqrt{4/R\delta}$.

If the longitudinal characteristic length of the surface imperfection is much larger than the depth the wake field amplitude is strongly reduced [9] and negligible compared to the resistive wake field.

Fig. 4 shows the wake potential for the LCLS parameter using a beam pipe radius of 2.5 mm. For larger radii, as it is the case for all helical undulators (A – C), the amplitude is noticeable reduced and the FEL performance is less degraded (Fig. 2). The wake fields are further reduced by using a low charge beam (case C), where the resulting effects are almost negligible. The benefit of a larger beam pipe is partially removed by a longer saturation length for case B because the energy modulation is accumulated over a longer distance.

The FEL amplification is less influenced in regions of the electron bunch where the total wake potential is zero. The average temporal radiation profiles for the LCLS case and case A are shown in Fig. 5. Again, the helical case exhibits superior performance

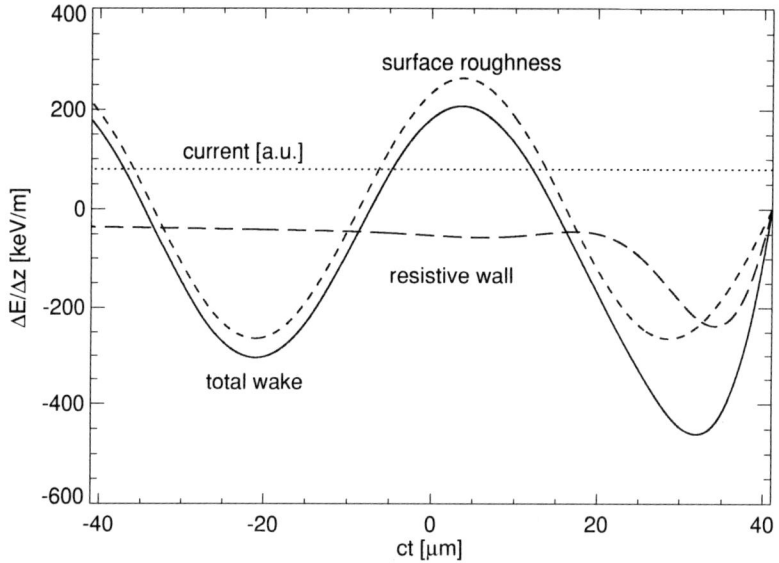

FIGURE 4. Wake potentials for the LCLS undulator

regarding the effective length of the radiation pulse, which is approximately 40% for case A but 10% for LCLS, compared to the electron bunch length.

CONCLUSION

The impact of spontaneous emission and wake fields on the X-ray FELs gain has to be taken into account during the design and manufacturing phase of the undulator. To reduce the amplitude of wake fields a larger chamber size and a shorter undulator are desired. This can be fulfilled by a novel designs of helical undulators with a very high magnetic field, leading to a consideration of these undulator for an X-ray FEL, instead of planar undulators based on a more conventional design. An additional benefit of helical undulators is the easier separation between spontaneous emission and FEL radiation, thus facilitating the diagnostic for the system. This feature shouldn't be underestimated during the commissioning phase of the FEL.

The energy loss by spontaneous emission can be compensated by field tapering and does not depend on the undulator type. Similarly the increase of the energy spread is mainly determined by the beam energy, and puts a limit to the shortest wavelength obtainable for SASE-FELs.

In conclusion our results show that the novel helical undulator design of the Kurchatov group, providing an unusual high magnetic field, has important advantages compared to planar undulators for an X-ray FEL.

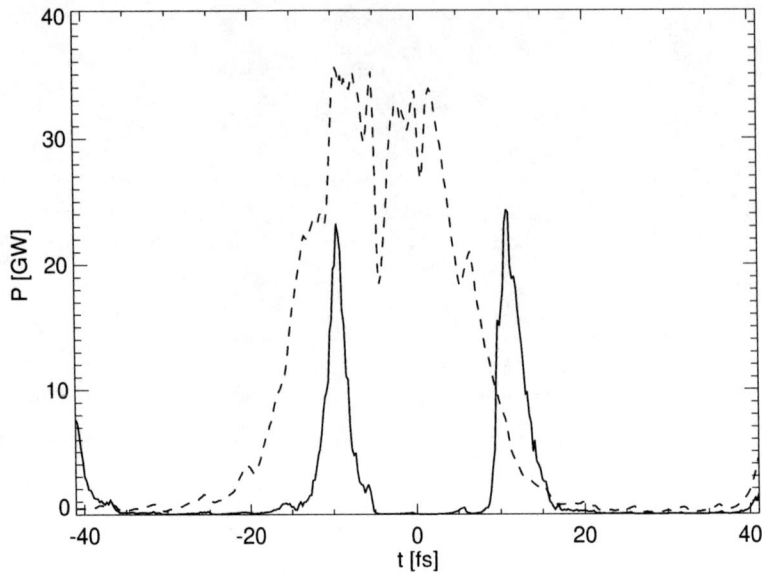

FIGURE 5. Average radiation envelope for the LCLS case and case A (solid and dashed line, respectively) close to saturation.

REFERENCES

1. LCLS Design Study Report, Report SLAC-R-521 (1998)
2. J.B. Rosenzweig and E. Colby, Proc. Conf. Advanced Acceleration Concepts, AIP**335** (1995) 724
3. S. Reiche, Nucl. Inst. & Meth. **A429** (1999) 243
4. S.G. Biedron *et al.*, Nucl. Inst. & Meth. **A445** (2000) 110
5. R. Bonifacio *et al.*, Phys. Rev. Letters **73** (1994) 70
6. E.L. Saldin *et al.*, DESY Print, TESLA-FEL, Hamburg, DESY, 1995
7. E.L. Saldin *et al.*, Nucl. Inst. & Meth. **A381** (1996) 545
8. A. Novokhatski and A. Mosier, Proc. of the PAC97 Conference, Vancouver, 1997
9. G. Stupakov *et al.*, Physical Review Special Topics – Accelerators & Beams **2** (1999) 060701
10. L. Palumbo *et al.*, Proc. of the EPAC00 Conference, Vienna, 2000
11. K. Bane, SLAC Report AP-87 (1991)

PARTICIPANTS LIST

Agafonov, Alexei V.
Lebedev Physical Institute, Russian Academy of Sciences, Nuclear Physics and Astrophysics, Leninsky prospect, 53, 117924 Moscow, Russia
agafonov@sci.lpi.ac.ru, phone: 095 132 6480, fax: 095 938 2251

Angelici, Marco
Univeristy of Rome "La Sapienza", via Euclide turba, 4, 00195 Roma, Italy
angelici@enrg58.ing2.uniroma1.it, phone: 39 06 322 3336, fax: 39 06 322 3173

Arthur, John
SLAC, SSRL, MS 69, 2575 Sand Hill Rd, P.O. Box 4349, Menlo Park, CA 94309-0210, USA
jarthur@slac.stanford.edu, phone: 650 926 3169, fax: 650 926 4100

Baldis, Hector
LLNL, Institute for Laser Science and Applications, MS L-411, 7000 East Avenue, Livermore, CA 94550, USA
baldis1@llnl.gov, phone: 925 422 0101, fax: 925 423 2463

Batterman, Boris
SLAC, SSRL, MS 69, 2575 Sand Hill Rd., Menlo Park, CA 94025, USA
bwb1@cornell.edu, phone: 650 926 8671, fax: 650 926 4100

Bazzani, Armando
Universita Di Bologna, Dipartimento Di Mathematica, Pizzo di Porta San Donato n. 5, 40127 Bologna, Italy
bazzani@bo.infn.it , phone: 39 51 351 095, fax: 39 51 247-244

Biedron, Sandra Gail
ANL, APS, B2164, 9700 S. Cass Ave., Argonne, IL 60439, USA
biedron@aps.anl.gov, phone: 630 252 1162, fax: 630 252 5703

Bionta, Richard M.
LLNL, Physics Directorate, N-Division, P.O. Box 808, L-050, Livermore, CA 94550, USA
bionta1@llnl.gov, stone9@llnl.gov, harris4@llnl.gov, phone: 925 423 4846, fax: 925 423 3371

Clendenin, James
SLAC, MS 66, 2575 Sand Hill Rd., P.O. Box 4349, Menlo Park, CA 94309, USA
clen@slac.stanford.edu, phone: 650 926 2962, fax: 650 926 2407

Cornacchia, Max
SLAC, SSRL, 2575 Sand Hill Rd., Menlo Park, CA 94025, USA
cornacchia@ssrl.slac.stanford.edu, phone: 650 926 3906, fax: 650 926 4100

Craft, Ben
LSU, CAMD, 6980 Jefferson Highway, Baton Rouge, LA 70806, USA
bcraft@mail.com, bcraft@unix1.sncc.lsu.edu, ph: 225 334 666, fax: 225 388 6954

Dattoli, Giuseppe
ENEA, FIS, Via Enrico Fermi, 45, 00044 Frascati (Roma), Italy
dattoli@frascati.enea.it, phone: 39 6 9400 5421, fax: 39 6 9400 5334

D'Auria, Gerardo
Sincrotrone Trieste, Accelerator Division, S.S.14 Km 163.5, Basovizza, I-34012 Trieste, Italy
gerardo.dauria@elettra.trieste.it, phone: 39 040 375 8220, fax: 39 040 375 8058

Emma, Paul
SLAC, NLC/LCLS, MS 66, P.O. Box 4349, Stanford, CA 94309, USA
emma@slac.stanford.edu, phone: 650 926 2458, fax: 650 926 5124

Fawley, William
LBNL, Accel. & Fusion Research Division, MS 71-J, 1 Cyclotron Rd, Berkeley, CA 94720, USA
fawley@lbl.gov, phone: 510 486 6229, fax: 510 495 2323

Fazio, Michael V.
LANL, LANSCE-9, MS H851, P.O. Box 1663, Los Alamos, NM 87545, USA
mfazio@lanl.gov, phone: 505 667 3281, 505 665 3568, fax: 505 667 8207

Ferrario, Massimo
INFN-LNF, Research Division, Via E. Fermi 40, C.P. 13, I-00044 Frascati (Roma), Italy
ferrario@lnf.infn.it, phone: 39 6 940 32216, fax: 39 6 32565

Floettmann, Klaus
DESY, MPY, Notkestr. 85, 22603 Hamburg, Germany
klaus.floettmann@desy.de

Freguglia, Paolo
University of L'Aquila, Department of Pure & Applied Mathematics, Via Panoramica 233,
I-58019 Porto S. Stefano GR, Italy
freguglia@univaq.it, phone: 39 05 648 18739, fax: 39 05 648 12298

Freund, Andreas
ESRF, Experimental Division, 6 rue Jules Horowitz, B.P. 220, 38043 Grenoble, France
freund@esrf.fr, phone: 33 47 688 2040, fax: 33 47 688 2542

Galayda, John
ANL, Accelerator Systems Division, Bldg. 401, 9700 Cass Avenue, Argonne, IL 60439, USA
deputy@aps.anl.gov, phone: 630 252 7796, fax: 630 252 7369

Gluskin, Efim
ANL, Experimental Facilities Division; Bldg 401, Rm. B3171, 9700 S. Cass Ave., Argonne, IL
60439, USA
gluskin@aps.anl.gov, phone: 630 252 4788, fax: 630 252 9303

Graves, William S.
BNL, NSLS, Bldg. 725D, Upton, NY 11973, USA
wsgraves@bnl.gov, phone: 631 344 2095, fax: 631 344 3238

Hajima, Ryoichi
JAERI, FEL Group, Tokai-mura, Ibaraki, 319-1195, Japan
hajima@popsvr.tokai.jaeri.go.jp, phone: 81 29 282 6315, fax: 81 29 282 6057

Huang, Zhirong
ANL, Accelerator Systems Division, Bldg 401, Rm. B2200, 9700 S. Cass Ave., Argonne, IL
60439, USA
zrh@aps.anl.gov, phone: 630 252 6023, fax: 630 252 5703

Jaeschke, Eberhard
BESSY, Machine Division, Rudower Chaussee 5, Geb 14.51, D-12489 Berlin, Germany
jaeschke@bessy.de, phone: 49 30 6392 4650, fax: 49 30 6392 4632

Kim, Kwang-Je
ANL, Accelerator Systems Division, Bldg. 401/C4265, 9700 S. Cass St., Argonne, IL 60439-
4800, USA
kwangje@anl.gov, phone: 630 252 4647, fax: 630 252 7369

Kincaid, Brian
LBNL, Accelerator & Fusion Research Division, 1 Cyclotron Rd., P.O. Box 12721, Berkeley, CA
94712, USA
bmkincaid@lbl.gov, phone: 510 486 4810, fax: 510 486 7066

Krejcik, Patrick
SLAC, Accelerator Department, MS 18, 2575 Sand Hill Rd., Menlo Park, CA 94025, USA
pkr@slac.stanford.edu, phone: 650 926 2790, fax: 650 926 3323

Lee, Richard W.
UCB, 101 Lombard Street #912W, San Francisco, CA 94111-1121, USA
dicklee@physics.berkeley.edu, phone: 415 421 0949, fax: 415 740 0821

Lee, Shyh-Yuan
Indiana University, Department of Physics, Swain Hall 117, Bloomington, IN 47405, USA
shylee@indiana.edu, phone: 812 855 2899, fax: 812 855 6645

Lelay, Guy
Orsay, France
lelay@crmc2.univ-mrs.fr

Liberto, Emanuele di
University of Rome "La Sapienza", Via Euclide Turba, 4, 0195 Roma, Italy
emadili@tiscalinet.it
Limberg, Torsten
DESY, MPY, Notkestr. 85, 22603 Hamburg, Germany
Torsten.Limberg@desy.de, phone: 49 40 8998 3998, fax: 49 40 8998 4305
Limborg, Cecile
SLAC, LCLS, P.O. Box 4349, Menlo Park, CA 94309, USA
limborg@SLAC.Stanford.edu
Lindau, Ingolf
SLAC, SSRL, MS 69, P.O. Box 4349, Stanford, CA 94309-0210, USA
lindau@ssrl.slac.stanford.edu, phone: 650 723 1052, fax: 650 926 4100
Materlik, Gerhard
DESY, HASY Lab, Notkestrasse 85, D-22607 Hamburg, Germany
materlik@desy.de, phone: 49 40 8998 2484, fax: 49 40 8998 4475
Matsushita, Tadashi
KEK, Institute of Materials Structure Science Photon Factory, Oho 1-1, Tsukuba, Ibaraki
305-0801, Japan
tadashi.matsushita@kek.jp, phone: 81 298 79 6020, fax: 81 298 64 3202
Mattioli, Mario
University of Rome "La Sapienza", Physics Department, Piazzale A. Moro, 2-00185 Roma, Italy
mario.mattioli@roma1.infn.it, phone: 39 06 499 14202
Migliorati, Mauro
Universit' di Roma "La Sapienza", INFN-LNF, Dipartimento di Energetica, Via A. Scarpa 14,
00161 Roma, Italy
migliorati@uniroma1.it, phone: 39 6 497 66343, fax: 39 6 442 4 0183
Milton, Stephen
ANL, Accelerator Systems Division, Bldg 401, Rm B2170, 9700 S. Cass Ave., Argonne, IL
60439, USA
milton@aps.anl.gov, phone: 630 252 9101, fax: 630 252 5703
Monteiro, Sergio
Moorpark College, Dept of Physics, Moorpark, CA 93021, USA
monteiro@physics.ucla.edu, phone: 805 378 1400, fax: 805 378 1499
Monteiro, Sergio
Moorpark College, 1325 Wellesley Ave #209, Los Angeles, CA 90025, USA
monteiro@physics.ucla.edu, phone: 310 442 9107
Ng, King Y.
FNAL, Beam Physics, P.O. Box 500, Batavia, IL 60510, USA
ng@fnal.gov, phone: 630 840 4597, fax: 630 840 6039
Nuhn, Heinz-Dieter
SLAC, SSRL, MS 69, P.O. Box 4349, Menlo Park, CA 94309-0210, USA
nuhn@slac.stanford.edu, phone: 650 926 2275, fax: 650 926 4100
Palumbo, Luigi
Universit' di Roma "La Sapienza", INFN-LNF, Dipartimento di Energetica, Via A. Scarpa 14,
00161 Roma, Italy
lpalumbo@lnf.infn.it, palumbo@axrma.uniroma1.it, phone: 39 6 497 66533, fax: 39 6 442 40183
Pellegrini, Claudio
UCLA, Dept of Physics & Astronomy, P.O. Box 951547, Los Angeles, CA 90095-1547, USA
pellegrini@physics.ucla.edu, phone: 310 206 1677, fax: 310 206 5251
Perfetti, Paolo
Italian National Research Council, ISM-CNR, Department of Physics, Via del Fosso del Cavaliere
100, 00133 Roma, Italy
perfetti@dns.ism.rm.cnr.it, phone: 39 06 49934477
Pflueger, Joachim
DESY, HASYLAB, Notkestr 85, 22603 Hamburg, Germany
Joachim.Pflueger@desy.de, phone: 49 40 8998 3242, fax: 49 40 8998 4475

Piot, Philippe
DESY, MPY, Building 30B, 85 Notkestrasse, 23606 Hamburg, Germany
philippe.piot@desy.de, phone: 49 040 8998 2755, fax: 49 040 8998 4305
Poole, Mike
CLRC Daresbury Laboratory, Synchrotron Radiation Department, Daresbury, Warrington, WA4
4AD, UK
m.w.poole@dl.ac.uk
Reiche, Sven
UCLA, Department of Physics & Astronomy, P.O. Box 951547, Los Angeles, CA 90095-1547,
USA
reiche@pbpl.physics.ucla.edu, phone: 310 206 5584, fax: 310 206 5251
Renieri, Alberto
ENEA, FIS, via Enrico Fermi, 45, I-00044 Frascati (Roma), Italy
renieri@frascati.enea.it, phone: 39 06 94005445, fax: + 9400 5607
Rosenzweig, James
UCLA, Dept of Physics & Astronomy, P.O. Box 951547, Los Angeles, CA 90095-1547, USA
rosen@physics.ucla.edu, phone: 310 206 4541, fax: 310 206 5251
Rossi, Carlo
Sincrotrone Trieste, Padriciano 99, 34012 Trieste, Italy
carlo.rossi@elettra.trieste.it
Rossbach, Joerg
DESY, MPY, Notkestrasse 85, D-22603 Hamburg, Germany
joerg.rossbach@desy.de, phone: 49 40 8998 3617, fax: 49 40 8998 4305
Ryynanen, Matti
Metallimiehenkuja 8, VTT Automation Mechnaics, P.O. Box 1303, FIN-2044 VTT, Finland
matti.ryynanen@vtt.fi, phone: 358 9 456 6485, fax: 358 9 455 3349
Shastri, Sarvjit
ANL, APS, 9700 S. Cass Ave., Argonne, IL 60439, USA
shastri@aps.anl.gov, phone: 630 252 0129, fax: 630 252 0161
Serafini, Luca
INFN-Milano, Via Celoria 16, 20133 Milan, Italy
luca.serafini@mi.infn.it, phone: 39 02 239 2673, fax: 39 02 239 2414
Schmerge, John
SLAC, SSRL, Blds. 137, MS 69, 2575 Sand Hill Rd., Menlo Park, CA 94025, USA
schmerge@slac.stanford.edu, phone: 650 926 2320, fax: 650 926 4100
Schmidt, Dominik
Stanford University, Electrical Engineering, 1618 Sand Hill Rd. #306, Palo Alto, CA 94304, USA
etruscan@stanford.edu, phone: 650 328 6642, fax: 650 328 6642
Sertore, Daniele
DESY, FDET, Notkestrasse 85, D-22607 Hamburg, Germany
daniele.sertore@desy.de, phone: 4940 8998 4907, fax: 40 40 8998 4448
Siddons, D. Peter
BNL, NSLS, Bldg 725D, Upton, NY 11973, USA
siddons@bnl.gov, phone: 631 344 2738, fax: 631 344 3238
Stohr, Joachim
SLAC, SSRL, MS 69, P.O. Box 4349, Stanford, CA 94309, USA
stohr@slac.stanford.edu, phone: 650 926 2570, fax: 650 926 4100
Stupakov, Gennady
SLAC, ARDA, P.O. Box 4349, Stanford, CA 94309, USA
stupakov@slac.stanford.edu, phone: 650 926 4320, fax: 650 926 5368
Sutton, Mark
McGill University, Dept of Physics, 3600 University St., Montreal QC H3A 2T8, Canada
sutton@physics.mcgill.ca
Tatchyn, Roman
SLAC, SSRL, 2575 Sand Hill Road, Menlo Park, CA 94025, USA
tatchyn@ssrl.slac.stanford.edu, phone: 650 926 2731, fax: 650 926 4100

Toor, Art
LLNL, Physics Division, L-047, P.O. Box 808, Livermore, CA 94550, USA
toor1@llnl.gov, stone9@llnl.gov, phone: 925 422 0953, fax: 925 423 0246

Tremaine, Aaron
UCLA, Department of Physics & Astronomy, Box 951547, Los Angeles, CA 90095-1547, USA
tremaine@pbpl.physics.ucla.edu, phone: 310 206 5584, fax: 310 206 5251

Tschentscher, Thomas
DESY, HASYLAB, Notkestr. 85, D-22607 Hamburg, Germany
thomas.tschentscher@desy.de, phone: 49 40 8998 3904, fax: 49 40 8998 4475

Uesaka, Mitsuru
University of Tokyo, Nuclear Enegineering Research Laboratory, 2-22 Shirakata-Shirane,
Tokai-mura, Naka-gun Ibaraki 319-1188, Japan
uesaka@tokai.t.u-tokyo.ac.jp, phone: 81 29 287 8421, fax: 81 29 287 8488

Weyer, Heinz J.
Paul-Scherrer Institut, SLS Project, CH-5232 Villigen PSI, Switzerland
heinz-josef.weyer@psi.ch, phone: 41 56 310 3494, fax: 41 56 310 3151

van der Wiel, Marnix J.
Eindhoven University of Technology, Department of Applied Physics, P.O. Box 513, 5600 MB
Eindhoven, The Netherlands
M.J.van.der.wiel@tue.nl, phone: 31 40 247 4321, fax: 31 40 243 8060

Varfolomeev, Alexander
Russian Research Center, Kurchatov Institute, 1 Kurchatov Square, Moscow 123182, Russia
varfolom@lki.msk.ru, fax: 70 95 196 77 64

Veneziano, Gabriele
CERN, CH-1211, Geneva 23, Switzerland
gabriele.veneziano@cern.ch

Walker, Richard
Sincrotrone Trieste, S.S. 14, Km. 163.5 Area Science Park, 34012 Basovizza, Trieste
richard.walker@elettra.trieste.it, phone: 39 040 375 8225, fax: 39 040 375 8565

Yu, Li-Hua
BNL, NSLS, Bldg 725C, Upton, NY 11793-5000, USA
bowden@bnl.gov, phone: 631 344 5169, fax: 631 344 3029

Yurkov, Mikhail
Joint Institute for Nuclear Research, LSVE, Dubna, 141980 Moscow Region, Russia
mikhail@lnf.infn.it, yurkov@mail.desy.de

AUTHOR INDEX

A

Agafonov, A. V., 59
Angelici, M., 33
Audebert, P., 45

B

Baldis, H. A., 45
Bazzani, A., 211
Biedron, S. G., 118, 203

C

Cauble, R. C., 45

D

Dattoli, G., 118, 203
Di Liberto, E., 107

F

Faatz, B., 162
Fateev, A. A., 162
Fawley, W. M., 118
Feldhaus, J., 162
Ferrario, M., 87
Floettmann, K., 162
Freguglia, P., 211
Freund, H. P., 118, 203
Frezza, F., 33, 107
Fronzoni, L., 211

G

Gauthier, J.-C., 45
Gluskin, E., 131

H

Hafz, N., 11
Hemker, R. G., 11
Huang, Z., 118, 185, 203

I

Ilinski, P., 131

K

Kim, K.-J., 185
Kinoshita, K., 11
Kobayashi, T., 11
Krzywinski, I., 162

L

Landen, O. L., 45
Lebedev, A. N., 59
Lee, R. W., 45
Lewellen, J. W., 118
Lewis, C., 45

M

Milton, S. V., 118, 203
Monteiro, S., 124
Mostacci, A., 33

N

Nakajima, K., 11
Ng, A., 45
Nuhn, H.-D., 23, 118, 203, 221

O

Okuda, H., 11
Ottaviani, P. L., 203

235

P

Palumbo, L., 33, 107
Pellegrini, C., 23, 221
Pflueger, J., 162

R

Reiche, S., 221
Renieri, A., 1, 203
Riley, D., 45
Rose, S. J., 45
Rosenzweig, J., 221
Rossbach, J., 162

S

Saldin, E. L., 153, 162, 169
Schneidmiller, E. A., 153, 162, 169
Schroeder, C. B., 23, 221
Serafini, L., 87
Shvyd'ko, Y. V., 153
Stupakov, G. V., 141
Sutton, M., 6

T

Tolmachev, S. V., 73, 221
Tschentscher, T., 162

Turchetti, G., 211

U

Ueda, T., 11
Uesaka, M., 11

V

Varfolomeev, A. A., 73, 78, 221
Vinokurov, N., 131

W

Wark, J. S., 45
Watanabe, T., 11

Y

Yarovoi, T. V., 78
Yoshii, K., 11
Yurkov, M. V., 153, 162, 169